W9-ATO-398

HOW TO GUIDE GIRL SCOUT JUNIORS ON

GET MOVING!

IT'S YOUR PLANET—LOVE IT! *A LEADERSHIP JOURNEY*

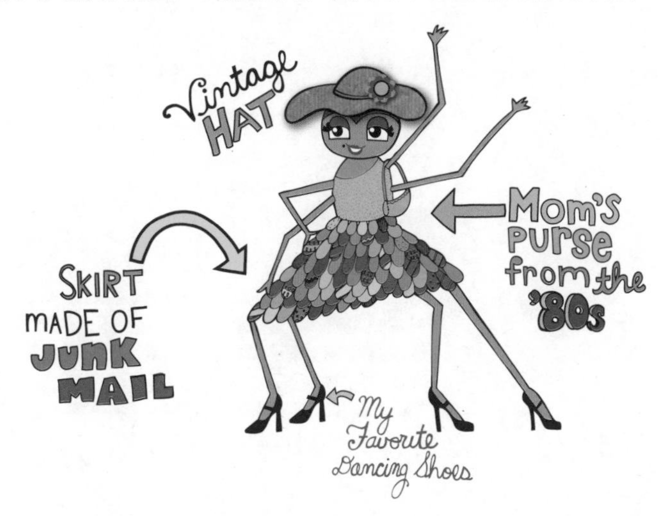

A Girl Scout leadership journey invites girls to explore a theme through many experiences and from many perspectives. All the joys of travel are built right in: meeting new people, exploring new things, making memories, gathering keepsakes. This guide is your suitcase. It's packed with everything you need for a wonderful trip that will change girls' lives.

Girl Scouts of the USA

CHAIR, NATIONAL BOARD OF DIRECTORS
Connie L. Lindsey

CHIEF EXECUTIVE OFFICER
Anna Maria Chávez

CHIEF OPERATING OFFICER
Jan Verhage

VICE PRESIDENT, PROGRAM
Eileen Doyle

PHOTOGRAPHS
Page 28: Photo of California Academy of Sciences courtesy of Charmagne Leung. **Page 41:** Beads by Valerie Takahama. **Page 85:** Fist-to-Five Consensus-Building originally adapted for *Agent of Change* (GSUSA, 2008) from Fletcher, A. (2002). FireStarter Youth Power Curriculum: Participant Guidebook. Olympia, Wash.: Freechild Project.

The women mentioned in this book are examples of how women have used their voice in the world. This doesn't mean that GSUSA (or you) will agree with everything they have ever done or said.

Text printed on Fedrigoni Cento 40 percent de-inked, post-consumer fibers and 60 percent secondary recycled fibers. Covers printed on Prisma artboard FSC Certified mixed sources.

SENIOR DIRECTOR, PROGRAM RESOURCES: Suzanne Harper

ART DIRECTOR: Douglas Bantz

WRITER: Valerie Takahama

CONTRIBUTORS: Toi James, Kate Gottlieb, Kathleen Sweeney, Cheryl Kushner, Amélie Cherlin

ILLUSTRATOR: Meghan Eplett

DESIGNER: Alexander Isley Inc.

EXECUTIVE EDITOR, JOURNEYS: Laura Tuchman

GSUSA DESIGN TEAM: Sarah Micklem, Rocco Alberico

© 2009 by Girl Scouts of the USA

First published in 2009 by Girl Scouts of the USA
420 Fifth Avenue, New York, NY 10018-2798
www.girlscouts.org

ISBN: 978-0-88441-739-2

All rights reserved. Except for pages intended for reuse by Girl Scout volunteers, this book may not be reproduced in whole or in part in any form or by any means, electronic or mechanical, including photocopying, recording, or by any information storage or retrieval system now known or hereafter invented, without the prior written permission of Girl Scouts of the United States of America.

Printed in Italy

5 6 7 8 9/17 16 15 14 13 12

This Girl Scout journey is funded in part by Ingersoll Rand.

STATEMENT OF TRUST

Girl Scouts of the USA creates national program materials to serve our vast and diverse community of girls. To help bring topics "off the page and into life," we sometimes provide girls—and their volunteers—with suggestions about what people across the country and around the world are doing, as well as movies, books, music, web pages, and more that might spark girl interest.

At Girl Scouts of the USA, we know that not every example or suggestion we provide will work for every girl, family, volunteer, or community.

In partnership with those who assist you with your Girl Scout group, including parents, faith groups, schools, and community organizations, we trust you to choose "real life topic experts" from your community, as well as movies, books, music, websites and other opportunities that are most appropriate for the girls in your area and that will enrich their Girl Scout activities.

Thank you for all you do to bring the Girl Scout Leadership Experience to life with girls, so that they become leaders in their own lives—and the future leaders the world needs!

CONTENTS

"Energy creates energy."

—Sarah Bernhardt, French actress, 1845–1923

Get Set to *GET MOVING!*

Our big, spinning world is teeming with energy—starting with the abundance of energy in your crew of Girl Scout Juniors!

GET MOVING! calls on girls across the country and around the world to use *their* energy to protect *Earth's* energy. The journey invites Juniors to engage their minds and hearts as they explore the many forms, uses, and misuses of energy. ***GET MOVING!*** challenges girls to safeguard Earth's precious energy resources by using their leadership skills—their ability to energize themselves and others, and their ability to investigate and innovate. Ultimately, they will educate and inspire others to use energy more efficiently, too!

Along the way, Juniors will tune in to the energy of the natural world. What they learn about energy and how revved up they feel when exploring the great outdoors will be the engine that drives them to care for our planet now and throughout their lives.

Juniors are joining an enduring tradition. The wonders of nature and the need to care for them have been at the core of Girl Scouting since its founding in 1912. As one early Girl Scout handbook noted: "The girl learns in Scouting that from a pebble to a star, all nature waits only to be explored in order to yield unending possibilities of fun and adventure, of resources waiting only for her mind and hands to appropriate and use."

That's the spirit of this journey, too: Energy is everywhere, and the possibilities are endless for its wise and innovative use. So what are you waiting for? ***GET MOVING!***

Energy and Leadership

To engage Juniors in the many aspects of energy, *GET MOVING!* makes use of the natural connection between energy and leadership. It's no coincidence that ENERGIZE, INVESTIGATE, and INNOVATE are the three prestigious awards girls can earn along this journey. Leaders are:

- **naturally energetic:** They keep themselves energized and know how to energize others.
- **natural investigators:** Their curiosity enables them to get to the root of issues and search out promising solutions.
- **innovative:** No matter what they tackle, they aim for fresh ideas and fresh approaches.

GET MOVING! engages Juniors as leaders who understand the importance of energy—their own and the planet's—and why it must be used wisely. The journey gives girls the confidence and skills they need to Take Action to conserve Earth's energy and to make a difference in any arena they choose.

HEARTS AND MINDS

So much information is now available about environmental problems and what must be done to correct them. *GET MOVING!* is part of a series of Girl Scout leadership journeys that invites girls, and their families and adult volunteers, to make sense of that information so they can act for the betterment of Earth.

The umbrella theme for the series—*It's Your Planet—Love It!*—came directly from a brainstorm with teen Girl Scouts. Its sentiment is clear: The desire to nurture and protect is first and foremost an act of love. If girls love the planet and all of its precious energy resources, they will naturally be moved to protect it. Love for Planet Earth is the true and necessary starting point for thoughtful and sustained environmental action.

PERFECT TIMING

What makes Girl Scout Juniors ready for the challenge of energized and energizing leadership? The timing is perfect because Juniors are just beginning to understand that they can make a difference in the world beyond their families, friends, and classrooms. They can talk and partner with others, expand their world and their worldview, and be part of an effort of change. Most important, they are at an age when they believe they have something to contribute to the world.

Like all Girl Scout leadership journeys, *GET MOVING!* reflects the Girl Scout leadership philosophy of **Discover** (understanding self and values), **Connect** (inspiring and teaming with others), and **Take Action** (acting to make the world a better place).

The energy that starts within each girl and then spirals outward is like a growing web. You'll notice references to webs throughout the girls' book, starting with the playful character of the fashion-conscious spider, Dez I. Ner—better known as Dez. Dez offers quips and commentary along the journey as she gains an understanding of her own over-the-top energy use and learns to unplug herself cord by cord to become more energy-efficient.

YOU AND THE GIRLS, TRAVELING TOGETHER

You may already be deeply committed to environmental causes—or not. Either way, you will be guiding girls on a journey of learning and doing that creates larger currents of energy in their lives and in the world. Along the way, the Juniors will enjoy what makes Girl Scouting unique among after-school offerings: Girl Led activities in a cooperative atmosphere of learning by doing.

SCIENCE, MATH, NATURE, AND YOU!

All along its energetic route, *GET MOVING!* engages girls in science, math, the outdoors, and environmental stewardship. You may be an expert in one or all of these areas—or none. No matter—there's no need to have all the answers. All you need to guide your group of Juniors is right here in this book. Just add your own energy, and an eagerness to explore all that energy offers and accomplishes in partnership with girls.

Imagine the power of more than 500,000 Juniors and their volunteers and families making choices that conserve and protect Earth's energy. What are you waiting for? *GET MOVING!*

A BOOK OF THEIR OWN

Keep in mind that the girls' book is designed so the Juniors can read it on their own at any time. It offers plenty for them to think about and do by themselves or in small groups—but they don't need to do everything. And it's theirs—to add to and cherish as they progress through all of life's journeys.

Awards Along the Journey

Get Moving! offers Juniors a chance to earn three prestigious Girl Scout leadership awards—Energize, Investigate, and Innovate. The girls can choose to earn one, two, or all three. If they earn all three, they'll see how the awards join together to create an energizing effect on their vest or sash.

To earn the **Energize Award**, girls:

- make an Energy Pledge to reduce their energy use in one or more ways
- try at least two other Energize activities suggested along the journey
- check out how other people are tackling energy issues

To earn the **Investigate Award**, girls connect with their Girl Scout crew to:

- learn about energy use in their buildings
- work with their families to make an energy improvement at home
- investigate energy use in a community building and suggest ways to make it more energy-efficient

To earn the **Innovate Award**, girls:

- identify an energy issue in the community, research it, create a plan, and carry it out, all the while reaching out to others to join in, too
- share the news, reflect on what they accomplished, and celebrate it

Tracking the Girls' Progress

The Energy Award Tracker that begins on page 106 of the girls' book lets the Juniors record their progress throughout the journey. Encourage them to fill it in as they complete each step toward an award. By journey's end, they'll see just how energized their Girl Scout team and the community can be.

Three Ceremonies or One Big One?

The girls may want to decide ahead of time whether to have award ceremonies as they earn each award or at the end, to celebrate earning them all.

Energy, Connections, and Commitment

In Girl Scouts, using energy wisely is often about acting thoughtfully to make something better. When girls join together to develop and encourage personal skills, they energize themselves both as individuals and as a group. They also gain a personal connection and commitment to their team, and this strengthens their involvement in whatever the team chooses to do.

WHAT IF A GIRL MISSES AN AWARD STEP?

Find a way for her to do something similar to what she missed so she can still earn the award with her team. If she misses the day that the team practices communication skills, she can partner with a girl and trade tips at a later time. If she misses out on the energy audit, she can take the lead in a follow-up step, such as reviewing the audit data and making recommendations for greater energy efficiency.

You might call on the full team of Juniors to brainstorm together about how girls who miss some steps can best get back on track with the journey.

When girls miss a team meeting, your goal is to assist them in finding ways to have a similar learning and growing opportunity—and to understand how they can contribute to the team. Girls may not have the exact same experience, but they can each take away new insights, connections, and a sense of accomplishment.

Snapshot of the Journey

SESSION 1 Start Your Engines	Juniors begin to experience the various forms of energy and how they can make the most of their own energy to conserve Earth's energy. The girls: • get an overview of the journey and its prestigious leadership awards • discuss the basics of energy and energy efficiency • make recycled paper • name their unique personal energy
SESSION 2 Pledging to Save Energy	Juniors commit to an energy pledge, take a look at how plants use energy, and sort through some of the wasted energy they see around them—all activities leading to the Energize Award. The girls: • share details of their energy pledges and consider a Team Energy Pledge • assess the waste involved in excess packaging • make beads from recycled paper • consider how leaders use energy
SESSION 3 Get Wild about Energy (and How to Conserve It)	Juniors investigate how animals use energy according to their needs and consider what humans might learn from them. They also compare and contrast animal and human communication strategies. The girls: • observe animals in their natural setting • consider how animal communication compares to human communication • create their own list of communication do's and don'ts
SESSION 4 Investigating Buildings	Juniors begin to explore energy use in buildings as they delve deeper into the science of energy and get ready to conduct an energy audit of a community building. The girls • test the energy efficiency of various lightbulbs • check their meeting space for drafts and brainstorm how to stop them
SESSIONS 5 & 6 The Energy Audit	The Juniors conduct an energy audit of a community building and then educate and inspire others about the importance of energy efficiency as they move toward their Investigate Award. The girls: • collect and review the energy data of their building • practice communication skills as they prepare to report their energy-efficiency suggestions to the building's officials

SESSION 7 Gearing Up to Go	The Juniors begin thinking about their Innovate project. The girls: • discuss the merits of various project ideas • learn interviewing techniques as they prepare to meet with energy experts • discuss their ideas for energizing food choices • take time to make silhouettes
SESSION 8 Moving in New Directions	The Juniors move toward a team decision on an Innovate project. Depending on the girls' interest, they may conduct a walkability/bikeability survey of their community. Then they: • make a team decision on their project • begin to plan next steps
SESSIONS 9 & 10 Innovate!	The Juniors plan and carry out their Innovate project, taking action to create changes in energy use on Earth and educating and inspiring others along the way. The girls also: • check in on their teamwork and conflict-resolution strategies
SESSION 11 Crossing the Finish Line	The Juniors reflect on and celebrate their accomplishments along the journey. The girls: • earn the Innovate Award • look ahead to more energizing adventures in Girl Scouting

Creating a Friends and Family Network

If other parents or creative teens can assist with any journey experiences, by all means enlist their help. You don't have to do it all! Use the Junior Friends and Family Network Welcome Letter and Checklist on pages 18–19 to enlarge your web of volunteers.

Health, Safety, and Well-Being

FIRE SAFETY

Young girls can be taught to start a camp fire under the watchful supervision of an adult. Be mindful of local bans on building fires and policies on open flames. It's always possible to enjoy a simulated, no-flame campfire!

Girl Scouting is guided by a positive philosophy of inclusion that benefits all. On this journey, it is hoped that girls will increase their feelings of being powerful, capable, and strong as they enhance their skills and develop new ones. So, as the Girl Scout Law says, "be a sister to every Girl Scout." Be sensitive to whether any girls are new to town, have a disability, don't speak English as a first language, or have parents getting a divorce. Often what counts most is being open-minded and aware, staying flexible, and creatively varying your approach with the girls.

The emotional and physical safety and well-being of girls is of paramount importance in Girl Scouting. Look out for the safety of girls by keeping *Volunteer Essentials* and the Girl Scout reference "Safety Activity Checkpoints" handy. And when planning all gatherings and trips, be sure to:

- Check into any additional safety guidelines your Girl Scout council might provide, based on local issues.
- Talk to girls and their families about special needs or concerns.

Welcoming Girls with Disabilities

First, don't assume that because a person has a disability, she needs assistance or special accommodations. Probably the most important thing you can do is to ask the individual girl or her parents or guardians what she needs to make her experience in Girl Scouts successful. If you are frank with and accessible to the girl and her parents, it's likely they will respond in kind, creating a better experience for all.

It's important for all girls to be rewarded based on their best efforts—not completion of a task. Give any girl the opportunity to do her best and she will. Sometimes that means changing a few rules or approaching an activity in a more creative way. Here are a few examples:

- Let a girl perform an activity after observing others doing it first.
- Let the girls come up with ideas on how to adapt an activity.

Often what counts most is staying flexible and varying your approach with the girls.

For a list of online resources, visit girlscouts.org and search on "disability resources."

GIRL SCOUT COUNCIL CONTACT INFO

Name:____________________

____

____

Can help with:____________

____

____

Phone:____________________

E-mail:____________________

Juniors and the Great Outdoors

CAMPING, ANYONE?

If you're worried about homesickness, be sure to:

- Pair "sleepover pros" with less experienced girls.
- Keep the girls active and busy.

NOT AN OUTDOOR PRO?

Ask for assistance from your Girl Scout council. Perhaps an experienced volunteer can join you when you venture out with the girls. Or consider obtaining camping certification for yourself. Councils offer it, often over a single weekend, to teach safety skills, orienteering tips, and other outdoor how-tos. Some adults describe the training as a retreat—an enjoyable opportunity to spend time with like-minded women. Contact your council to learn more.

Outdoor experiences broaden a girl's perspective and build an appreciation for the natural world that can lead to a lifetime of environmental stewardship.

For Junior-age girls, outdoor adventures also promote courage, strength, and open-mindedness. Learning how to follow a trail, put up a tent, or prepare a meal in the wild builds confidence, too—and teamwork. Hikes and other outings, no matter how short or close to home, will likely inspire a desire for bigger adventures in the great outdoors.

So as *GET MOVING!* unfolds, make time for excursions into nature. Let the girls choose their outings based on their interests and skills, and the availability of volunteers in your network who can assist them.

Whenever possible, let outdoor excursions be a time for the girls to think about all the energy around them—and ways to reduce their use of Earth's energy, whether for transportation or other personal needs. Camping is a great energizing and energy-reducing option. An "off the grid" weekend might be the ideal way for the girls to test their Energy Pledges and their endurance in honoring them.

Light at Night

Observing the night sky for 30 minutes to an hour makes for an enjoyable nighttime outing. The following "Story Book in the Sky" section, adapted from the *Junior Girl Scout Handbook* of 1963, is a good way to get girls in a night sky mood.

STORY BOOK IN THE SKY

Long ago people told stories in the evening after the day's work was done. They looked up at the dark skies and named groups of stars after the heroes and animals and objects in their stories. These groups of stars are called constellations, which means "stars together." When looking for constellations, the Big Dipper is a good one to start with because the handle and edges of the dipper point to other constellations.

Did you know that you can get directions from the North Star and the sun if you do not have a compass? At night you can see the North Star in the Little Dipper. When you face the North Star, you are facing north.

Get the girls talking about what they observe in the sky by asking:

- *Did you read the constellation story on page 58 of your book? What did you think of it?*
- *How about making up our own story about a constellation?*

INVESTIGATING LIGHT POLLUTION

While enjoying the night sky, keep the energy theme going by encouraging the girls to take stock of the quality of the night's darkness. Talk with the girls about the area's level of light pollution, which is unnecessary light brightening the night and diminishing the natural view of the night sky.

Light pollution can come from many sources, including city signs, building lights, and inefficient street lights that send light out and up rather than just down. As the girls view the night sky in their community, you might ask: *In what ways is light pollution a problem or not a problem here?*

If the amount of artificial light in the night sky is high, the girls might choose to address light pollution in their Innovate project. They may even want to add in a full evening gathering to observe the night lights.

NIGHT SIGHTS

The girls might want to choose a park or a preserve to look at the constellations. Let them know that sky charts are available online or at a local library to guide them in viewing the night sky. And don't forget to bring along flashlights or a lantern.

Call on your Junior Friends and Family Network to see if anyone has a telescope to lend or might otherwise assist the Juniors. If an outdoor excursion isn't possible, perhaps the girls can visit a planetarium.

Lure of the Campfire

Fire is an enduring symbol of energy, and campfires have an uncanny ability to draw out the storyteller in all of us. So make the most of these energizing gatherings. If you and the girls are in a "no-fire zone," improvise with a pyramid of flashlights and colored tissue paper.

If fires are permitted in your camping area, consider making homemade firestarters from recycled materials. You'll need: cardboard egg cartons, dryer lint, ends of burned-down candles. (Ask the girls to begin collecting these items a few weeks before you plan to make your campfire.)

To make 12 firestarters, fill the egg carton compartments halfway with dryer lint. Melt candle ends in double boiler, then pour the wax into each compartment until full. When ready to use, place one "egg" in the kindling and light a match.

PLAYING WITH SHADOWS

Shadow puppetry might begin spontaneously as lantern light plays on the walls of the girls' tents. Consider taking it up a notch by combining it with some energizing stories.

What stories might the Juniors like to tell? What characters might they create? A variety of flashlights will accentuate the drama!

Shadow puppets can be cut from recycled cardboard and taped to sticks. Or the girls can just use their hands.

Girl Scout Stories of Getting Outdoors

It was so cold that the dishwashing soap froze and our matches got wet. Luckily we had already started our campfire. Ms. Trueblood showed us how to save the hot coals by covering them with ash and dirt. In the morning, we scraped off the cool ash and dirt and found that some of the coals from the previous fire were still hot. Ms. Trueblood took some wadded brown paper sack and some dry twigs and started that fire again without any matches.

—Jennifer Beaty-Elbert, Lawton, Oklahoma

WOW, a real campfire, I couldn't wait! I still remember the moon's reflection on the lake in front of me and the warm glow of my friends' faces by the fire. I learned so many new campfire Girl Scout songs that night. I remember seeing my first shooting star . . . and also finding a piece of meteorite. I also had another first that night, a S'more. Mmmmm.

—Melissa Madsen, Stamford, Connecticut

Girl Scout Traditions and Ceremonies

For nearly 100 years, Girl Scout traditions and ceremonies have connected girls to one another, to their sister Girl Scouts and Girl Guides around the world, and to generations of Girl Scouts who came before them.

A few traditions are mentioned here; your council will have many more. Try incorporating them into Girl Scout gatherings and get-togethers. And be sure to involve the girls in creating and passing on new traditions.

THE GIRL SCOUT SIGN

The Girl Scout sign is made when saying the Girl Scout Promise. It is formed by holding down the thumb and little finger on the right hand and leaving the three middle fingers extended (these three fingers represent the three parts of the Promise).

THE GIRL SCOUT HANDSHAKE

The Girl Scout handshake is the way many Girl Guides and Girl Scouts greet each other. They shake their left hands while making the Girl Scout sign with their right hand. The left-handed handshake represents friendship because the left hand is closer to the heart than the right.

WIDENING THE FRIENDSHIP CIRCLE

The Friendship Circle and friendship squeeze is often formed as a closing ceremony for meetings or campfires. Everyone gathers in a circle, crosses their right arm over their left, and holds hands with the people on either side. Once everyone is silent, one girl starts the friendship squeeze by squeezing the hand of the person to the left. One by one, each girl passes on the squeeze until it travels around the circle.

Because this journey emphasizes a love of nature and Earth, you might widen the Juniors' circle from time to time by inviting teens or adults to one of your opening or closing ceremonies to share their outdoor experiences or what they've done for the good of the planet.

PLAN FOR AN ENERGIZING FINALE

Early on, invite the girls to start thinking about how they'd like to celebrate the culmination of this journey together. And along the way, invite them to create as many smaller celebrations as they like. Each time the Juniors gather can be an energizing occasion.

For ceremony ideas, visit girlscouts.org and search on ceremonies.

Welcome!

Your Girl Scout Junior has joined a team of girls on a journey all about energy. As the girls explore their own energy as leaders and the energy of Planet Earth, they'll give special focus to the energy of places and spaces and the energy of getting from here to there. Along the way, they will have the chance to earn three prestigious leadership awards: Energize, Investigate, and Innovate.

As a team, the Juniors will plan a project that gets others moving in ways that are good for Earth. This Innovate project has the potential to have a lasting effect on the community.

Your guidance and expertise can help make the girls' experience along *GET MOVING!* even more valuable and memorable. Please take a moment to review the attached checklist to let us know which areas you might have expertise in, or time to volunteer for, so that your Junior and her sister Girl Scouts will have the richest experience possible. Then please help your Junior and her team by identifying areas in which you feel you could contribute time or talent—for the benefit of all Juniors.

The girls and I look forward to hearing from you—and seeing you at sessions throughout the journey.

Sincerely,

____

Junior *GET MOVING* Guide

Contact Info

Phone __________________________________

E-mail __________________________________

Girl Scout council (name and phone number) ____

Yes, I want to assist the Juniors as they *GET MOVING!*

I am ready to volunteer by offering:

- ☐ ideas about improving the way people get from here to there (transportation)
- ☐ knowledge of energy use in buildings
- ☐ assistance with science experiments
- ☐ several small houseplants for use in the girls' plant energy experiment
- ☐ art, building, or craft skills
- ☐ practical experience with the outdoors
- ☐ time as a driver (if needed for outings)
- ☐ time as an all-around Junior leadership journey helper
- ☐ to bring energizing snacks to the session

My Girl Scout Junior is: ______________________________

My name: ______________________________

My contact info: ______________________________

____

____

____

YES! I know of community energy issues and energy experts that might interest the Juniors as they Investigate and Innovate.

1. ______________________________
2. ______________________________
3. ______________________________
4. ______________________________

What + How: Creating a Quality Experience

It's not just what girls do but how you engage them that creates a high-quality experience. All Girl Scout activities are built on three processes that make Girl Scouting unique from school and other extracurricular activities. When used together, these processes—Girl Led, Learning by Doing, and Cooperative Learning— ensure the quality and promote the fun and friendship so integral to Girl Scouting. Take some time to understand these processes and how to use them with Juniors.

Girl Led

"Girl Led" is just what it sounds like—girls play an active part in figuring out the what, where, when, how, and why of their activities. So encourage them to lead the planning, decision-making, learning, and fun as much as possible. This ensures that girls are engaged in their learning and experience leadership opportunities as they prepare to become active participants in their local and global communities. With Juniors, you could:

- encourage girls to plan and lead a session, activity, project, or event
- model and provide strategies for solving problems and making decisions
- expose girls to diverse ideas, geographies, and cultures, noting the similarities and differences from their own

On this journey the girls take the lead in Session 2, when they share examples of wasteful packaging they've found in their own homes. They then have the opportunity to contrast that wasteful packaging with the ancient, and energy-efficient Japanese packaging known as *furoshiki*—before they go on to design their own smart and energy-efficient packaging style.

KEEP IT GIRL LED

Remember: You want the girls to take a major role in planning and executing this leadership experience. The girls may first want you to come up with the ideas and plans. *But hold your ground!* This is the girls' experience, and they're up to the challenge.

From beginning to end, keep your eye on what the girls want to do and the direction they seem to be taking. It's the approach begun by Juliette Gordon Low: When she and her associates couldn't decide on a new direction, she often said, "Let's ask the girls!"

At each session, ask the girls for any last thoughts on what they've done or discussed.

Learning by Doing

Learning by Doing is a hands-on learning process that engages girls in continuous cycles of action and reflection that result in deeper understanding of concepts and mastery of practical skills. As they participate in meaningful activities and then reflect on them, girls get to explore their own questions, discover answers, gain new skills, and share ideas and observations with others. Throughout the process, it's important for girls to be able to connect their experiences to their lives and apply what they have learned to their future experiences both within and outside of Girl Scouting. With Juniors, you could:

- talk with the girls about ways to connect their learning to daily life
- guide girls to reflect on their learning experiences by using the many ideas in this journey
- support girls' hands-on testing of their own ideas, skill-building, and teaching skills

REFLECTING ON HOW FAR THEY CAN GO

In Session 3, the girls reflect on their Energy Pledges and consider how they might go even further "off the grid" in their own lives as they encourage others to join with them in conserving energy.

Cooperative Learning

Through cooperative learning, girls work together toward shared goals in an atmosphere of respect and collaboration that encourages the sharing of skills, knowledge, and learning. Given that many girls desire to connect with others, cooperative learning may be a particularly meaningful and enjoyable way to engage girls in new ideas and knowledge. Working together in all-girl environments also encourages girls to feel powerful and emotionally and physically safe, and it allows them to experience a sense of belonging even in the most diverse groups. With Juniors, you could:

- structure experiences so that girls "need" each other to complete the task
- use role-play and realistic scenarios to guide girls in communicating and working effectively within groups
- give girls examples of how individuals manage their cooperative groups, such as assigning roles, assessing how they are doing, and staying on task

TALKING THEIR WAY THROUGH A MAZE

In Session 5 or 6, the girls have an opportunity to pair up to complete a maze as a way of understanding the importance of clear communication among team members. That's Cooperative Learning, Learning by Doing, and Girl Led all rolled into one!

Seeing Processes and Outcomes Play Out in *GET MOVING!*

Girl Scout processes and leadership outcomes play out in a variety of ways during team gatherings, but often they are so seamless you might not notice them. For example, in Session 3 (page 44), the Juniors go out into nature to observe animals in their natural habitat. The call-outs below show how the Girl Scout processes and outcomes make this activity a learning and growing experience for girls—and up the fun, too! Throughout *GET MOVING!*, you'll see processes and outcomes play out again and again. Before you know it, you'll be using these valuable aspects of Girl Scouting in whatever Juniors do—from planning a camping trip to earning the Girl Scout Bronze Award.

FROM SAMPLE SESSION 3

Get Wild about Energy (and How to Conserve It)

Opening Ceremony

This is an example of the **Girl Led** process. Opening ceremonies should always be Girl Led, with girls taking the lead on conducting the ceremony. At the Junior grade level, the girls can determine how to modify ceremonies to suit their interests.

Since they're gathering at a special place to observe animals in nature, the girls might form a circle and name one hope they have for their outing today.

Observing Animals in Their Natural Setting

This is a good example of **Learning by Doing.** Girls are going out in nature and observing animals for themselves, rather than watching someone else do it or just reading about it. There is an added element of complexity and benefit to the girls when they record their observations and thoughts, perhaps for later use. This could also relate to the Discover outcome, **Girls seek challenges in the world,** since the Juniors are going out and exploring the world and their thoughts and ideas.

If you and the girls have arranged to get out in nature to observe animals, this is the time to do it! Encourage the girls to jot their thoughts, notes, and drawings of all that they see and hear and smell.

If your group isn't out in nature, viewing a nature movie or some nature shows on TV (or taped from TV) are a good options. They can observe the animals and then follow the same discussion points and activities.

This section is an excellent example of the adult-girl partnership in the **Girl Led** process. The adult volunteer is guiding the girls in a discussion by asking them pointed questions about their experiences and observations. The "W and H questions" (who, what, when, where, why, how) are especially useful for getting the ball rolling for a discussion with younger girls.

Animals, Energy, and Movement

After the girls have finished their allotted time for observing animals, guide them in a discussion about what they've seen. Here are some questions you might ask to get them started:

- *What sort of animal energy and animal movement did you see today?*
- *How is movement necessary to an animal's survival?*
- *How do animals use sound energy?*
- *How do humans use the energy of animals?*
- *Did you read about the phrase "charismatic megafauna" from the story about the scientist who studies elephants in your book? Why do you think we like certain animals enough to try to protect them?*

While this question is part of the "W questions" noted above, it also serves another purpose: moving the girls toward achieving the Discover outcome, **Girls develop critical thinking**. It is asking girls to take information they have gathered and use it to explain a complex idea, such as the reasoning behind protecting animals.

PETS ARE ANIMALS, TOO

Steer the girls into a related discussion about energy and the animals they're likely to be most familiar with: pets. To get them going, pose a few questions, such as:

- *Do cats or dogs or other pets get as much chance as deer or prairie dogs or other wild animals to run around and move freely?*
- *What happens when pets don't get to use their energy properly?*
- *How can people make sure pets get enough exercise?*
- *Do you notice how exercising a pet gets you some exercise, too?*

In addition to being a "W question" to get girls thinking about the health of pets, this could also be the preliminary question for moving toward the Take Action outcome, **Girls are resourceful problem solvers**. It's asking girls to gather their existing resources to find a solution to the problem of pets needing more exercise.

This last question, though it requires only a yes or no answer, gets girls to start thinking about different ways to stay healthy, which is the Discover outcome, **Girls gain practical life skills**.

Understanding the Journey's Leadership Benefits

Filled with fun and friendship, this journey is designed to develop the skills and values girls need to be leaders in their own lives and as they grow. Activities along *Get Moving!* are designed to enable girls to achieve 12 of the 15 national outcomes of the Girl Scout Leadership Experience, as detailed on the next page.

Each girl is different, so don't expect them all to exhibit the same signs to indicate what they are learning along the journey. What matters is that you are guiding the Juniors toward leadership skills and qualities they can use right now—and all their lives.

For full definitions of the outcomes and the signs that Girl Scout Juniors are achieving them, see *Transforming Leadership: Focusing on Outcomes of the New Girl Scout Leadership Experience* (GSUSA, 2008). Keep in mind that the intended benefits to girls are the cumulative result of traveling through an entire journey—and everything else girls experience in Girl Scouting.

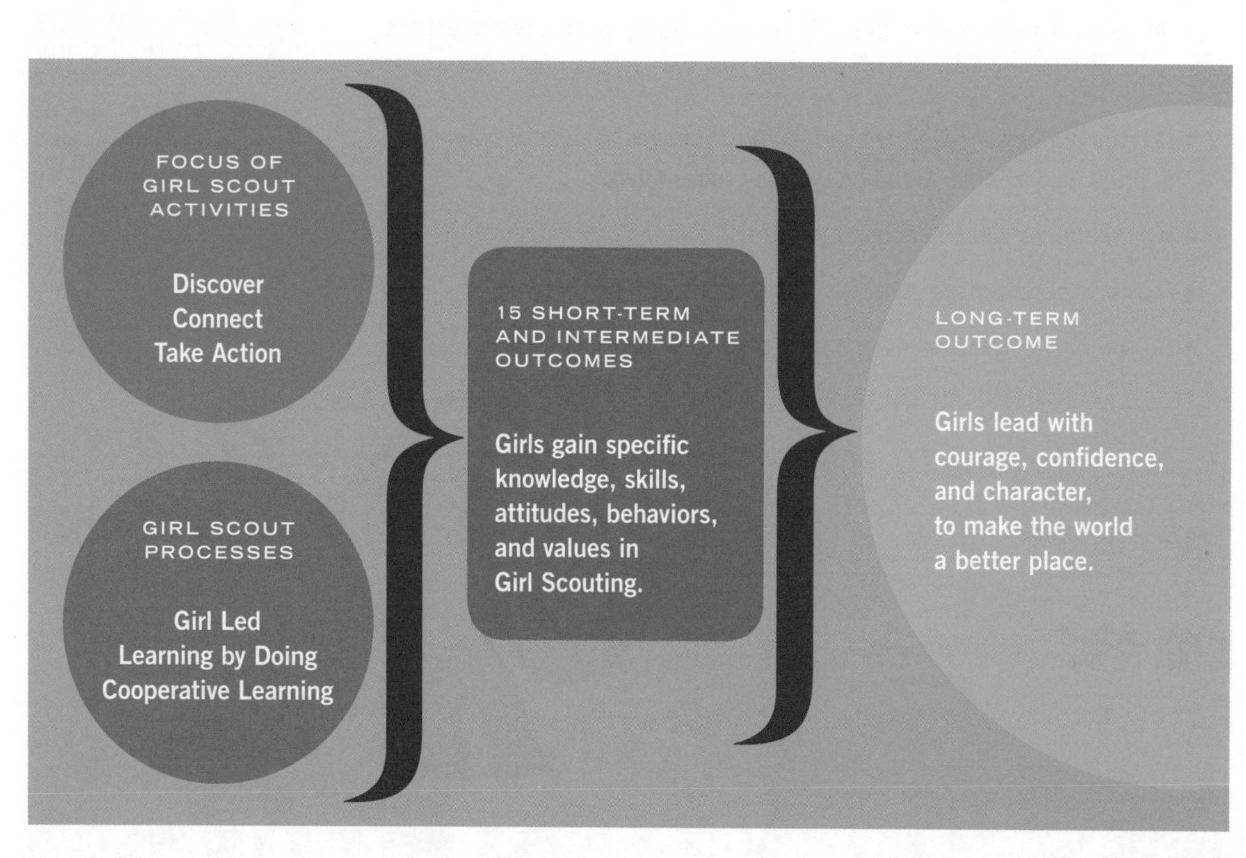

NATIONAL LEADERSHIP OUTCOMES

		AT THE JUNIOR LEVEL, girls...	RELATED ACTIVITIES (by Session number or girls' book part/page)	SAMPLE "SIGN" When the outcome is achieved, girls might...
DISCOVER	**Girls develop a strong sense of self.**	gain a clearer sense of their individual identities.	S1: Personal Energy	report increased confidence in dealing with outside pressures.
	Girls develop positive values.	gain greater understanding of ethical decision-making in their lives.	S1: Making Recycled Paper; GB: Energy-Saving Adventure, p. 21, Recycled Paper, pp. 36-39	give examples of using the Girl Scout Promise and Law in deciding to "do what's right."
		have increased commitment to engage in sustainable community service and action.	S2 and S4: Opening Ceremonies; Carbon Footprint, pp. 16–17; Energy Pledge, pp. 18–20; Weighing Waste, p. 32; Energy Award, p. 106; Moving Right Along, pp. 111–112	feel it's important to help people and the environment in ways that will have a long-term positive impact.
	Girls gain practical life skills.	gain greater understanding of what it means to be emotionally and physically healthy.	Energy Snacks, adult guide, p. 30; GB: Any Bean Soup, p. 23; Walking Salad, p. 31	describe how being stressed can affect physical health.
	Girls seek challenges in the world.	increasingly recognize that positive risk-taking is important to personal growth and leadership.	S2: Energy Pledge; S4: Opening Ceremony; S5 & S6: Communicate with Style; S7: Lightbulbs	when asked to identify attitudes important to accomplishing goals, mention risk-taking and give examples from their own lives
		are better at exploring new skills and ideas.	S1: Recycled Paper; S2: Plants, Light; S3: Observing Animals, Animal Energy Movement; S4; S5 & S6: Energy Audit; GB: Kinetic and Potential Energy, pp. 10–12; You Unplugged, p. 14–17; Wasted Energy, pp. 44–49; Energy Insights, pp. 52-59; Recycled Paper, pp. 36–39; Energy Audit, pp. 68–75; On the Move, pp. 76-79; Investigate Award, p. 107	report using a variety of resources to pursue topics of interest.
CONNECT	**Girls promote cooperation and team-building.**	increasingly recognize how cooperation contributes to a project's success.	S5 and S6: Communication Maze; S7: Thinking About a Team Choice	consistently prefer solving problems in teams or as a group and explain why this can be more effective than working alone.
	Girls feel connected to their communities.	begin to feel part of a larger community of girls/women.	GB: Profiles of women/girls, pp. 22, 24-25, 34-35, 40, 41, 42-43, 50, 51, 67, 83, 84	enjoy connecting with girls/women locally, nationally, or globally.
	Girls develop healthy relationships.	strengthen communication skills for maintaining healthy relationships.	S3: Relate and Communicate!; S5 & S6: Relate and Communicate: Favorite Tips	name communication strategies that help them in their relationships.
TAKE ACTION	**Girls can identify community needs.**	learn to use strategies to determine issues that deserve action.	S8: Walkability/Biking Checklist	use community asset mapping to identify opportunities to better their communities.
	Girls are resourceful problem solvers.	are better able to create an "action plan" for their projects.	S5 and S6: How Does Your Building Stack Up?; S8: Innovate Project; S9 & S10: Planning Time: Innovate; GB: An Innovative Idea, pp. 80-81; Innovate Award, pp. 108–109	outline steps, resources, and time lines and assign responsibilities for their project with minimal adult guidance.
		gain a greater ability to locate and use resources that will help accomplish their project goals.	S5 and S6: Energy Audit; S7: How to Conduct an Interview; GB: An Innovative Idea, p. 35; Innovate Award, pp. 108–109	feel confident contacting community partners who can help them achieve their goals.
	Girls advocate for themselves and others.	strengthen their abilities to effectively speak out or act for themselves and others.	S4–S10: Investigate and Innovate projects; S5 & S6: Speak Up for Change; GB: Investigate, p. 107	identify concrete steps they can take to effect desired changes.
	Girls educate and inspire others to act.	learn various strategies to communicate and share Take Action Projects with others.	S2: How Leaders Energize; S5 & S6: Speak Up for Change; S11: Celebrate	use various ways to tell others about their Take Action Projects.
			S2: Team Strand of Beads; S3: Communication Dos and Don'ts; S5 & S6: Communicate with Style	explain what makes a successful persuasive message/action for various audiences.
	Girls feel empowered to make a difference.	are more confident in their power to effect positive change.	S7: Lightbulbs	describe various expressions of power around them.
		feel they have greater opportunities for involvement in the decision-making of their communities.	S11 Reflecting on the Journey; GB: Innovate Award, pp. 107–108; Celebrate, p. 112	explain how shared power helped them create better or longer-lasting changes.

S=Session, GB=Girls' Book

From *Agent of Change* to *GET MOVING!*

DON'T GIVE AWAY THE POWER!

If *GET MOVING!* is the first Girl Scout leadership journey your Junior team is embarking on, skip these "power" tips. You wouldn't want to give away the fun and friendship of the *Agent of Change* before the girls have a chance to uncover it on their own! Just go ahead and enjoy *GET MOVING!*

If your Junior Team has already enjoyed the *Agent of Change* journey, keep those powerful experiences growing by linking some of its leadership ideas to the energy of *GET MOVING!* You might talk to the girls about how their *GET MOVING!* experiences also let them experience ever-widening circles of power. For example:

- When the girls explore their own personal energy, they are delving into the Power of One.
- When they team up to investigate a building and then plan an Innovate project, they are experiencing the Power of Team.
- And when they reach out to wider circles to educate and inspire others with their Innovate project, they experience the Power of Community.

As the Juniors travel through *GET MOVING!*, take a moment from time to time to ask them questions that help them make larger links to this growing web of power. You might ask:

As you move from exploring personal energy to investigating the energy of places and spaces to delving into the energy of getting from here to there, how is your web of power strengthening and expanding? As your network widens, do you see how much more you can accomplish?

Your Perspective on Leadership

The Girl Scout Leadership philosophy—Discover + Connect + Take Action—implies that leadership happens from the inside out. It stresses the importance of embracing who you are, connecting with others, and working collaboratively to make things better for all. Throughout this journey, you and the girls will have a richer experience if you use your own reflections as part of how you inspire the girls. Take time now—and throughout *GET MOVING!*—to apply the three "keys" of leadership to yourself.

Discover + Connect + Take Action = Leadership

DISCOVER **What value do you have that motivates you to protect Earth? How will you share this with the girls? What does the Girl Scout Law line "use resources wisely" mean to you?**

____

____

____

CONNECT **What is the most energizing experience you have ever had while on a team? How can you guide Juniors to have this kind of experience?**

____

____

____

TAKE ACTION **How does your role as a volunteer with Girl Scout Juniors contribute to making the world—and specifically the environment—better?**

____

____

____

> "We're seeing everything from . . . recycling . . . to zero-energy homes . . . taking hold . . . across the nation. I'm asking you to keep pushing. I'm daring you to out-green each other."
>
> —Lisa P. Jackson, administrator, Environmental Protection Agency, speaking to the Local Government Advisory Committee, Washington, D.C., March 2009

Green roof, California Academy of Sciences, San Francisco

The Journey's 11 Sample Sessions

Think of the session plans in this guide as a basic road map—they'll take you and the girls through each step of the journey, accomplishing goals along the way. Each session is designed to allow the girls to make choices, so that the journey is as meaningful and memorable as possible.

Check out all the options for customizing the journey and adding in energizing snacks on pages 30–31. As you and the girls make this journey your own, keep in mind that the use and misuse of energy is the main theme.

Encourage the girls to think wisely about resources from the get-go. Here are some tips to get you started for your first gathering:

Session 1, beginning on page 32, gets the girls into some crafty fun with a papermaking activity. You might:

- Get in touch with the girls ahead of time and invite them to collect recycled or natural decorative bits—ribbons, threads, fabric, postage stamps, wrapping paper, anything they might be able to find to make their paper unique. (You might also tap the Friends and Family Network to supply decorative bits.)

- Encourage the girls to collect fallen leaves, blades of grass left from lawn mowing, or fallen flower petals to use in their paper.

- If the location of your gathering is conducive to it, the girls might collect fallen leaves and flowers petals just before the activity. That combines using resources wisely with another big journey theme: getting outdoors to enjoy nature! (Just remember to emphasize Leave No Trace principles: Collect what has fallen to the ground, but don't disturb living things!)

Customizing the Journey

If you and your "travelers" have time, be imaginative. Ask the girls how and when they'd like to get "off the highway" to stop at their own "roadside attractions." They might want to add on:

Story stuff: Opportunities to experience how stories of all kinds are created and shared—whether comics (maybe an artist can visit with the girls?), plays, films, books (writing, illustration, and design), photo essays, art exhibits, oral histories.

Trips to energizing spots in the girls' communities: Whether visiting a waterfall or an amusement park, encourage the girls to jot their observations. Sometimes the most innovative ideas are right in their own backyard—the nearest playground, library, or school. Even a trip to the mall can spark the girls' imaginations.

Outdoor (or indoor) activities that offer plenty of physical activity: If not enough are suggested in this guide, create some of your own—even if they're not elaborate.

Ask the girls what they enjoy—classics like tag, time to go a little crazy with freeze-dancing to music, a jumping-jacks break? And ask them where they like to go. Their favorite places may be good spots for outdoor breaks, and they may be places that the girls will revisit for their Innovate project.

They might also just want to go outdoors for a walk, a picnic, or to practice map reading and compass skills.

Crafts: If girls express interest, encourage crafts opportunities as they arise. After the Juniors complete their Energy Pledges, some girls may want to make more recycled paper beads to trade with one another—it keeps the energy theme going. Craft projects can also become mementos that the girls will keep or incorporate into ceremonies.

NEED TO SCALE BACK? JUST STAY FOCUSED!

If you are limited to 6–8 sessions, keep in mind that you don't have to do every activity. You'll get the most out of each one by keeping it small in scale and focused. If possible, allow for Sessions 5 & 6 to be longer so that the girls have the time they need to engage in their energy audit.

KEEP IT GIRL LED

Let any ideas that the girls latch on to take them in the direction *they'd* like to go. For instance, the girls may want to extend their observation time in nature, or add on a hike or camping trip and then visit some animal-filled cityspaces.

Energy Snack Suggestions

Energizing snacks made of healthful foods are effortless to prepare. Consider offering a variety of them throughout the journey. Invite the girls to take the lead in choosing and preparing them. You might get them started with these suggestions:

Lettuce "envelopes": These "wraps" offer a fun tie-in to the letter-writing that takes place at various points along the journey, as the girls invite guest speakers and reach out to legislators. Lettuce can be flattened and rolled up—just like paper! Place a few chopped tomatoes, torn basil leaves, and cubes or strips of mozzarella cheese (or a cheese of your choice) on a lettuce leaf, and then roll! The girls might try these simple veggie snacks as part of their Energy Pledge, which is a step toward the ENERGIZE award.

Fresh fruit like blueberries, raspberries, and cherries are a healthful snack. Plus, they're bead-like! Consider serving them during Session 2, when the girls make paper beads from magazines or other colorful "used" paper.

Trail mix is a good snack for outdoor activities. The girls might enjoy it during Session 3, when they're out viewing animals in nature.

Hot apple juice with a cinnamon stick or another hot drink might be an energizing break for Session 4, when the girls begin to investigate energy use in buildings—especially if it's a cold day with a lot of chilly drafts!

Orange slices or "smiles": Brainstorming requires a lot of energy, and citrus fruits like oranges are among the foods with the best "brain sugars"—that is, those that don't cause radical ups and downs in blood sugar levels. So consider them as a snack for Session 7 or 8, as the girls dig in on their Innovate ideas.

Popcorn lightbulbs: Use your favorite recipe for popcorn balls, but instead of forming them into round balls, shape them into lightbulbs!

Make-your-own granola bars: Combine rolled oats and wheat germ with sweeteners like honey and brown sugar. Then customize with seeds, nuts, and dried fruits of the girls' choice.

GET READY TO GET ENERGIZED

During Session 1, you'll get the girls thinking about what food is really energizing to them. If you need to brush up on nutrition tips, visit MyPyramid.gov, the food pyramid Web site of the U.S. Department of Agriculture.

Sample Session 1

Start Your Engines

This opening session makes use of the first two sections of the girls' book, "Off the Charts, Off the Grid" and You, Unplugged," which offer information on:

- *The definition of energy and the basic types of energy, pages 10–11*
- *The Energy Pledge the girls will commit to, pages 19–21*
- *How much energy is used in the making of paper, page 26–29*
- *Papermaking (pages 36–39) and artist Akua Lezli Hope, who uses bits of nature in her paper (page 40)*

AT A GLANCE

Goal: Juniors begin to experience the various forms of energy and how they can make the most of their own energy to conserve Earth's energy.

- **Opening Ceremony: Energy Is Everywhere!**
- **Awards Along the Journey**
- **Making Recycled Paper**
- **Energizing Guests**
- **Energizing Snacks**
- **Closing Ceremony: Personal Energy**
- **Looking Ahead to Session 2**

MATERIALS

- **Making Recycled Paper:** A piece of screen; recycled paper: tissue paper, newsprint, printer paper, wrapping paper, envelopes, ripped into small pieces (about one-half inch square), recycled or natural decorative bits (optional); plastic basin for water run-off; electric blender; rags (lots) and a few old towels; newspapers; butter knife; pieces of cardboard cut to match the size of the papermaking screen or frame; duct tape or other heavy-duty tape.
- **Closing Ceremony:** Poster board or chalkboard (or get creative with recycled materials) and colored markers.

PREPARE AHEAD

Suggest that the Juniors wear old clothes or smocks for the papermaking.

Also tap the Friends and Family Network to supply decorative bits for papermaking and the ingredients for any snack the girls might make (see suggestions, page 31).

Opening Ceremony: Energy Is Everywhere!

Invite the girls to form a Friendship Circle and introduce the theme of the journey: Energy and its wise use!

- Let the girls know that scientists define energy as the ability to do work. It's that simple! Say something like: *Energy can be found everywhere! In one form or another, energy is connected to almost everything we do.*

Ask the girls to think of all the ways they've used energy or seen energy being used in the past 24 hours. You might ask: *Did you take a shower, play with the cat, or text a friend? Did you toss a ball? Figure out a math problem?*

- Kick off the conversation by sharing one way you used energy recently. You might say something like: I had friends over last night, so I took a dessert from the freezer and zapped it in the microwave to defrost it. Then we ate it. All of that takes energy.
- Then invite each girl to share her own energy example with the group. If any girls are stuck, try prompts such as, Did you read or watch TV last night? How did you get to school today?

Make a game of it. See if the girls can go around the circle two more times naming ways they've used energy.

Be sure to let the girls know that pages 10–11 of their books offer many interesting examples of all the forms of energy in the world. Encourage them to dip into those pages throughout the journey.

Next, ask the girls if they know the terms "energy efficiency" or "energy-efficient" and can explain what they mean. If no one knows, say something like: *Energy-efficient means using less energy to do the work needed.*

Now, ask the girls to go around the circle again and each name a more efficient use of energy for one of the energy uses they first named. You might lead off by saying, *Remember the dessert I served? What if I had taken it out of the freezer in time for it to thaw on its own? Then I wouldn't have had to use a microwave to defrost it. That would have been more energy-efficient.*

GRAND CEREMONIES ARE GIRL LED!

During this opening session, encourage the girls to start thinking about creating their own opening and closing ceremonies—ones that focus on energy. Invite the girls to volunteer to lead the ceremonies for future gatherings. You might even get a signup list started.

Awards Along the Journey

Introduce the girls to the three prestigious leadership awards they can earn along the journey. You might point them to page 7 in their book, and to the Energy Award Tracker on pages 106–109. Here's a handy way to summarize how the awards tie to leadership:

Energize: *To earn this award, you will explore what personal energy is, and all the ways you can use your personal energy to its fullest. Leaders are full of energy! They know how to energize themselves and everyone around them! That's what this journey is all about.*

Investigate: *As a team, you will investigate energy at home and in a building in the community. You'll come up with recommendations for how to improve its energy efficiency. Investigating is something all good leaders do! Leaders investigate issues and ideas in order to accomplish their goals.*

Innovate: *As a team, you will create and put in place a plan for how your community can conserve energy and use energy wisely. Great leaders are innovators! They come up with ways to do things better. But they don't do it on their own. They take in opinions from everyone around them in order to achieve the best results.*

Let the girls know that the steps to these awards are built into their team gatherings. All along the journey, they'll be developing the leadership skills and the energy smarts they need to earn their awards in a smooth and engaging way. As they *Get Moving!* they will also be moving forward with lasting change. Ask: *Why is it important to create positive change that lasts? What values of the Girl Scout Law encourage us to do that?*

Now's a great time to get the girls excited about their Energy Pledge. Ask them to take a minute together now to check out pages 19–21 of their book. What interests them the most about the sample pledge they see? What do they hope to include in their own pledges?

Making Recycled Paper

Recycling is a great way to save energy, so you might introduce this activity by saying: *There are lots of ways to turn old stuff into new stuff, and making new paper out of used paper is one of them.*

Making recycled paper allows the plant fibers of the original paper to be used over and over again. It also uses less electricity and less water, and creates a lot less pollution than making new paper. By making recycled paper as they will today, the girls also use their own energy in the papermaking! Using their own energy helps save Earth's energy! That's something they will notice again and again on this journey!

Ask: *By making our own paper and saving trees, what part of the Girl Scout law we are living?* (*Answer: Using resources wisely!*)

You might point the girls to the profile of papermaking artist Akua Lezli Hope (page 40 of their book).

Ask for a team of three or four volunteers to make the first piece, following the illustrated directions on pages 36–38 of their book. Other teams can make more sheets as time and materials allow. The sheets will need to dry thoroughly overnight.

PAPER FOR THE JOURNEY

Next, get the girls talking how they might use their recycled paper throughout the journey. The girls' suggestions might include:

- To write parts of our energy pledges at our next gathering.
- To create personal energy certificates for one another based on a personal quality each girl brings to the team.
- To make mini-posters with notes and pictures about special memories, energizing places we want to visit, or whatever strikes our fancy.
- To make special Energy Pledge charts or collages to remind us of our promises to save energy.
- To write letters inviting guests to our gatherings and to write thank-you notes to all who have helped us along the way.

WHAT SHOULD PAPER PULP LOOK LIKE?

The pulp the girls create should look like clouds and have the consistency of soupy oatmeal.

Energizing Guests

Ask the girls to think about special guests they would like to invite to talk with them about energy at various points in their journey. It might be especially helpful to look ahead at Sessions 5 and 6, which focus on an energy audit of a community building. Your council may be able to put you in touch with energy experts who will assist the girls.

Remember, energy connects us to so many people and issues. Encourage the girls to think about all the women and girls featured in their book and all the energy experts in their region who might be interesting to meet: those knowledgable about solar, wind, or other nonpolluting energy sources, engineers designing new forms of transportation, zoologists studying the movements of animals, astronomers concerned about city lights dimming the night sky, experts in making buildings energy-efficient, energizing chefs. The possibilities are endless!

Energizing Snacks

Ask girls to call out the advertising slogans they know for favorite foods or drinks. Jot key words on a large piece of paper or a chalk board. Now ask girls to rate how "energizing" these foods are, by marking them "E" for Energy or "D" for Doubt It (no matter how much we may enjoy it). After they've rated them, ask: *How do these advertisements influence your decisions about choosing energizing foods?*

Wrap up by asking the girls what kinds of energizing snacks they'd like to try each time they gather. Point them to the recipes in their book (pages 23 and 31) and share the snack suggestions on page 31 of this guide.

BE ON THE LOOKOUT FOR EXPERTS

You might want to tap into the Friends and Family Network for suggestions of energy experts and others to invite to a Junior gathering.

Network members may even have expertise on an energy subject or issue to share with the girls.

Consider partnering with other Junior groups to hold joint gatherings when special guests visit. That way, all the girls benefit!

Be sure to look for energizing guests who can share tips and recipes for energizing food choices.

INVITES MADE NICE

Talk with the girls about thoughtful ways to invite guests to gatherings.

You might say, *A nice invitation might be a personal letter.* Suggest that in a later gathering, once they've selected a guest, the girls might team up to write the invitation. And they can write it on their handmade recycled paper!

When the girls are ready to write a letter, make use of the letter-writing tips and bloopers on page 74, and the sample letters on page 75.

Closing Ceremony: Personal Energy

Invite the girls to sit in a circle. Explain that energy is all around us—it's within us, too: *Each person has her own unique personal energy and it's something she can use to be a good leader. Some people are bright, bouncy, or bubbly. Others are calm, focused, or steady.*

- Invite the girls to think of a word and image that describes their unique energy. Ask them to go around the circle and say the word aloud and how they might use that energy to be a good leader.
- Then each girl will draw an image of her energy on the poster board.

For example, one girl might say, "My energy word is 'busy.' I use my busy energy to accomplish my goals as a leader." Then she might draw a bee or a hummingbird to represent "busy."

- Go around the circle until each girl has had a chance to speak and draw.

Revisit this ceremony at future gatherings. Use the same poster board and see how the girls' feelings about their personal energy change over time. You might ask: *Have you found a type of personal energy that you enjoy the most and want to make the most of and be known for? A kind of energy that you want to carry with you always?*

Looking Ahead to Session 2

- Encourage the girls to arrive at the next gathering with specific ideas for their Energy Pledges. Encourage them to also create a Team Energy Pledge that covers how they use energy each time they gather.
- Ask the girls to save some colorful paper trash, such as magazines or supermarket ads or wrapping paper. Let them know that they'll use this for a fun bead-making activity at their next gathering.
- Also, ask the girls to save one piece of packaging, maybe something from "Weighing a Week of Waste" (see page 32 of their book), that they will use to play a guessing game at the next gathering. Encourage the girls to tear or cut the packaging so that it can't be easily identified. The idea will be for the rest of the Juniors to guess what the packaging once held.

Sample Session 2
Pledging to Save Energy

This session makes use of the second and third sections of the girls' book, "You, UnPlugged" and "Waste, Energy, and Wasted Energy," which feature information on:

- *Energy Pledges (page 19–21)*
- *Teen Girl Scout Abbe Hamilton (page 34)*
- *Recycle Cindy and Nancy Judd (pages 41–43)*

AT A GLANCE

Goal: The Juniors commit to an energy pledge, take a look at how plants use energy, and sort through some of the wasted energy they see around them.

- **Opening Ceremony: Good for Us, Good for the Planet**
- **Option: How About a Team Energy Pledge?**
- **The Packaging Puzzle**
- **Plants, Light, and Energy: An Ongoing Experiment**
- **Beads of Recycled Paper**
- **How Leaders Energize**
- **Closing Ceremony: Waste No More!**
- **Looking Ahead to Session 3**

MATERIALS

- **Opening Ceremony:** Poster board or chalkboard for writing Team Energy Pledge.
- **Plants, Light, and Energy:** Two or three house plants, paper bags. Paper and pencils.
- **Beads of Recycled Paper:** Paper from magazines, used wrapping papers, etc. Also rulers, pencils, and scissors; white glue; round toothpicks; elastic thread or soft-flex wire; optional: colorful seed beads or other beads, and a closure if using soft-flex wire.
- **Packaging Puzzle:** Depending on which way girls enjoy this activity, you may need a variety of packaging materials. Tap the Friends and Family Network for assistance.

PREPARE AHEAD

- Check out your meeting space to see if you can leave the plants from the Plants, Light, and Energy activity there. Or arrange for girls to care for the plants between sessions, following the needs of the experiment.
- Chat with any assistants about their roles before and during the session.

AS GIRLS ARRIVE

Invite the girls to set up the plants and bags for the Plants, Light, and Energy experiment, and the bead-making materials.

Opening Ceremony: Good for Us, Good for the Planet

Invite the girls into a circle to share a key point of the personal Energy Pledge they plan to commit to on this journey. You might say:

- *Keep in mind that using your own energy is good for you (energizing!) and good for the planet. When you use your own energy, you don't need to rely so much on Earth's energy!*
- *Remember that your pledge will involve key leadership skills: You're being a leader who energizes, investigates, and innovates!*

As each girl names her pledge point, have her write it on the poster board or chalkboard to create a collage of Energy Pledges. Then invite the girls to review all the pledge points together. Get a discussion going about how well these pledges work for the individual girls. You might pose questions like:

- *Do these pledges represent who you are as individuals and the best you can do as leaders who care about energy use? Are everyone's ideas represented? Is everyone's participation important? Why or why not? How do you know that your pledge is the best for you?*
- *Which of your ideas involve living the values of the Girl Scout Law?*
- *Is committing to your pledge going to be a challenge? Is it an example of positive risk-taking? How so?*
- *Why is taking risks important? Do your pledges offer different challenges for each of you? What might you learn from the challenge? From risk-taking in general?*

RECYCLED WRAPS: THE FUN OF FUROSHIKI

Furoshiki are cloths traditionally used in Japan to wrap and carry lunches, gifts, books, clothes, and other items. The custom dates back hundreds of years. It fell out of favor in the 20th century with the widespread use of plastic bags, but it's now undergoing a renaissance.

In 2006, former Japanese Minister of the Environment Yuriko Koike created a special *furoshiki* with a fabric made from recycled plastic bottles. She called it a *mottainai furoshiki,* because the Japanese word *mottainai* means "it's a shame to waste something."

Furoshiki are most commonly 18-inch and 28-inch squares. They are made from a variety of fabrics, such as rayon, cotton, silk, and nylon. There are special folding patterns to wrap objects of various sizes and shapes. They're ingenious and stylish! You can find directions online by simply searching on "furoshiki."

Share this energy-efficient tradition with the girls as part of the "Packaging Puzzle" activity. They might enjoy trying some, especially to carry a gift to younger girls. That way, the gift comes packaged in an energy lesson. Now, that's leadership!

Option: How About a Team Energy Pledge?

Now, if the girls are interested, get them really diving in on how they might go all out with a Team Energy Pledge. Ask: *How can we maximize our own energy use every time we meet in order to minimize the use of Earth's energy? How can we really test ourselves at every gathering along this journey, even if just for a short time? Should we:*

- *Go camping and stay off the grid for a whole weekend?*
- *Have a sleepover that is totally off the grid?*
- *Create some energy-saving "rules" to follow each time we meet?*

Encourage the girls to keep in mind that the more they use their own energy (to think, plan, do), the less of Earth's energy they'll need to use.
You might say:

- *That's good for us and good for the planet!*
- *What we're doing with our energy pledges is energizing ourselves and investigating how we can live with less of Earth's energy.*
- *And we're being innovative at the same time!*
- Ask: *How can we keep our innovation going for the long haul?*

The Packaging Puzzle

Invite the girls to sit in a circle to share the packaging items they've brought from home. The group has three chances to guess what product or food came in each package before the girl who brought it reveals its identity. After all the girls have shared and the group has finished guessing, get a discussion going about what they've just seen. You might ask:

- *Were you surprised at how much paper trash your family had available to choose from?*
- *Can you think of ways to reduce the amount of paper and other materials used in the packaging you saved or the packaging you've just seen?*
- *What was the most ridiculous or wasteful packaging you found?*

Then invite the girls to dig into the packaging materials you and the Friends and Family Network have provided, so they can put together their own smart, energy-efficient package or sketch their idea for it.

Plants, Light, and Energy: An Ongoing Experiment

Plants turn the sun's energy, and air and water, into stored energy in the form of sugars and starches. This experiment, which the girls will observe over the next several sessions, demonstrates how important sunlight is to plants.

Pass out paper and drawing utensils. Using the two or three small house plants that you have brought to the session, invite the girls to give one plant a normal growing spot either outdoors in the sun or indoors in a sunny area. This is the control plant. Then invite them to place the second plant next to the first plant, but ask them to deprive it of light by placing a bag or box that won't let light in over it for most of the day. This is the test plant. If the girls are using a third plant, it will serve as a second test plant. Ask them to cover the third plant for half of the available daylight hours. Let the girls know they must give each plant enough water to keep the soil a little moist.

Then pass out paper and drawing utensils and invite the girls to sketch all the plants, paying close attention to the leaves—what they look like and how many there are. You might ask: *Notice how all the leaves are healthy and green. As we watch these plants over the next few weeks, we're going to see if they change at all or continue to look just as they do right now.*

Beads of Recycled Paper

This activity uses triangles of paper cut from magazines, newspapers, wrapping paper, and the like to create delicate and attractive beads that can be fashioned into any jewelry items the girls desire. Let the girls know that their beads, a purely "recycled" creation, are a way to symbolize all the things they're doing to save energy. If they're game, they might even write some lines of their Energy Pledges or promises on their triangles of paper before they turn them into beads. That way, the words they know are hidden in their beads can serve as a reminder of all they can do to save energy.

BEADS NOT YOUR THING?

If the girls prefer, they can take a few scraps of paper, or strips of the recycled paper they made, and write ways for the team to be energy-efficient at its gatherings. Then invite the girls to read their suggestions, fold up their paper, and drop it in a jar or bowl that can be dipped into from time to time to remind the girls of energy-saving actions.

OPTION: A TEAM STRAND OF BEADS

In addition to the necklaces or bracelets the girls keep and wear or give to each other, they might also create a team strand. Each girl contributes a few beads representing the key parts of the energy pledge she will carry out. At the end of the journey, the Juniors can pass along this team strand to a group of girls who will be on the *Get Moving!* journey next and share its meaning with them. The Junior team might give their strand to a team of Brownies bridging to Juniors.

BEAD SWAPS

If the girls like, they can use leftover beads (or make a few extra) to make swaps. All they need to do is slide a bead onto a safety pin, and combine two or three of those pins onto a single pin.

Swaps ("Special Whatcamacallits Affectionately Pinned Somewhere") are a Girl Scout tradition of trading keepsakes as a way of making new friends and marking those friendships. The practice began years ago when Girl Scouts and Girl Guides first gathered for fun, song, and getting to know one another. Each Swap can spark a memory of a special event or an encounter with a sister Girl Scout.

To make the beads: One triangle of paper is used for each bead. The triangles must be long and thin—about 11 inches (or the length of a piece of paper) and 1 inch wide at their base. Here's a little geometry lesson to offer the girls: *Since two sides of each triangle are the same length, these are isosceles triangles.*

To make a bead from magazine pages: Tear out a page from a magazine; then, using a ruler and pencil, mark off 1-inch intervals along the bottom of the page. At the top of the page, place a mark at every half-inch. Now, place the ruler at the bottom edge of the page, angle it up to the 1-half inch mark, and draw a line. That's one long side of the triangle. Now, angle the ruler down to the first 1-inch mark at the bottom of the page and draw another line. That's the other long side of the triangle. Continue across the page to form more long triangles. Then cut the triangles apart with scissors.

To form a bead, take one of the triangles and dab glue lightly on one side of it, leaving a bit at the wide end uncoated. Wrap the paper, starting with the unglued bit, around the toothpick and roll the paper onto itself. Then push the bead off the toothpick.

Let the beads dry for 15 minutes. Then use elastic thread or other types of cord to make a necklace or bracelet. Between each paper bead the girls might intersperse seed beads, if desired, and then knot the ends.

How Leaders Energize

Now get the girls talking again about their Energy Pledges. Begin by asking:

- *What do you think will be the easiest parts of your pledge? The hardest?*
- *What about your pledge demonstrates your values? Your leadership? Our desire to create lasting change?*
- *Do you think you will be able to influence anyone else in your family to change the way they use energy? Not quite sure?*

You might say: *Remember, leaders energize—themselves and others! So think about how you might energize others so much that they join you in your Energy Pledge and make their commitment last, too.*

Ask: *Why is it important to create lasting change? What values of the Girl Scout Law tell us to aim for lasting change?*

Point out the story of Abbe Hamilton, the teen Girl Scout who got her school cafeteria to replace Styrofoam with compostable paper products (page 34 in the girls' book). Ask: *Could you imagine working together as a team to do something that creates lasting change at your school?*

Ask: *Why is teamwork and leadership important when you want to make changes in groups like a family, a school, or a community?*

Closing Ceremony: Waste No More!

Point the girls to the stories about Recycle Cindy and Nancy Judd, the Recycle Runway artist, on pages 41–43 of their book. You might say:

- *Many people take trash and find a way to make it into something new and innovative—and often beautiful!*
- *Let's each share one item we throw away that we'd like to find a new use for. This is one small step to saving energy.*
- *Once we* Get Moving! *on something small like this, we can inspire others to join in with us! That's leadership! How can we make the efforts of others last?*

Looking Ahead to Session 3

Chat with the girls about plans for their next gathering, when they'll spend time outdoors to observe how animals, plants, and insects use energy. Ask: *What place might be best for animal observing? Keep in mind we'd like to spend up to an hour there. Let's make the best team decision we can!*

As you guide the girls to make a team decision about where to go, based on their interests and accessibility, you might offer these suggestions: a park, a preserve, part of a local zoo, a dog park (at a dog park, the girls can observe how both animals and people use energy! Isn't it interesting how when we exercise our pets, we get exercise, too?).

- If the girls want to get in some exercise, they might plan a nature walk or a short hike and take in the animal life along the way. They'll be using their own energy as they observe the energy around them!
- If getting outdoors is not possible, the girls could visit an animal rescue center and talk with experts there who might have animals the team can observe.

As the team considers possible choices, ask: *Are we hearing everyone's thoughts? Is everyone feeling good about our decision? Why is that important to our teamwork? What might we do better next time we need to make a team choice?*

Sample Session 3

Get Wild About Energy (and How to Conserve It)

AT A GLANCE

Goal: Girls investigate how animals use energy according to their needs and consider what humans might learn from them. They also compare and contrast animal and human communication strategies.

- **Opening Ceremony**
- **Observing Animals in Their Natural Setting**
- **Animals, Energy, and Movement**
- **Relate and Communicate!**
- **Closing Ceremony**
- **Looking Ahead to Session 4**

MATERIALS

- **Observing Animals in Their Natural Setting:** Paper and pencils, energy snack of the girls' choice.
- **Relate and Communicate!:** Paper and pencils; Do and Don't template written on chalkboard or recycled poster board or paper.

PREPARE AHEAD

Check on any needed transportation and reach out to the Friends and Family Network for volunteers. After up to an hour of observing, encourage the girls to write or draw their observations and share them. So have paper and pencils handy!

Opening Ceremony

Since they're gathering at a special place to observe animals in nature, the girls might form a circle and name one hope they have for their outing today.

Observing Animals in Their Natural Setting

If you and the girls have arranged to get out in nature to observe animals, this is the time to do it! Encourage the girls to jot their thoughts, notes, and drawings of all that they see and hear and smell.

If your group isn't out in nature, viewing a nature movie or some nature shows on TV (or taped from TV) are a good option. They can observe the animals and then follow the same discussion points and activities.

Animals, Energy, and Movement

After the girls have finished their allotted time for observing animals, guide them in a discussion about what they've seen. Here are some questions you might ask to get them started:

- *What sort of animal energy and animal movement did you see today?*
- *How is movement necessary to an animal's survival?*
- *How do animals use sound energy?*
- *How do humans use the energy of animals?*
- *Did you read about the phrase "charismatic megafauna" from the story about the scientist who studies elephants in your book? Why do you think we like certain animals enough to try to protect them?*

PETS ARE ANIMALS, TOO

Steer the girls into a related discussion about energy and the animals they're likely to be most familiar with: pets. To get them going, pose a few questions, such as:

- *Do cats or dogs or other pets get as much chance as deer or prairie dogs or other wild animals to run around and move freely?*
- *What happens when pets don't get to use their energy properly?*
- *How can people make sure pets get enough exercise? Do you notice how exercising a pet gets you some exercise, too?*

ENERGIZE FOR ACTION

The Juniors might get a jump start on their Innovate project by reading and discussing "¡Vamos Ya," the comic story in their book (pages 85–101).

Suggest a group read, and encourage the girls to "take on" the roles of the fictional Juniors and the narrator.

Then ask:

- How do animals in the story use energy? And the girls? Is there a connection?
- The trip to the zoo leads to an unexpected event. What do the girls' reactions reveal about how they get along? What can be learned from the relationships in the story?
- How do the girls organize their project? What about their way of teaming up is good? What's not good?
- What about the girls' plan could help you energize your community to create lasting change?

KEEP RELATING

Capture the girls' tips and keep them handy at all gatherings. Check in on them with the girls from time to time. How is the team doing with them? Are they trying them out in other places in life, too? Where? How did it go?

ANIMALS AS INSPIRATION

Pass out the paper and pencils, and transition the girls to creating some images or poems inspired by what they've observed or talked about today. Encourage them to share what they've created with the group. To get their creative juices flowing, offer a few suggestions, such as:

- *Write a haiku (a poem, often in three lines of 5, 7, and 5 syllables, based on imagery from nature).*
- *Compose a rhyming chant or rap, or a silly song about how a creature looks, moves, or sounds, or about how you feel about how energy flows through nature.*
- *Sketch a picture of animal or insect using or creating energy.*

Relate and Communicate!

Next, get the girls thinking about how animals get along. What did they notice during their nature observations? What do they notice about their pets.

They might enjoy knowing that elephants may communicate with one another by stomping around, but ring-tailed lemurs (like those studied by Mireya Mayor, the primatologist profiled on page 50 of their book) have a quieter way of sending messages to their pals: They communicate through smell. They have scent glands on their wrists and chests that they use to mark trees in their territory. Lemurs also get into "stink fights" in which they rub their scent glands on their tails and then flick them at their enemies. A nicer way they communicate is by keeping their tails up like flags when they travel in groups. That way everyone stays together, kind of like a pack of traveling tourists on vacation!

You might ask:

- *How is the behavior of the elephants and the lemurs like that of humans?*
- *When are people likely to stomp around? When are they likely to use scents?*
- *Have you ever noticed groups of tourists in your area? Have you ever seen a group in which everyone is wearing the same hat, T-shirt, or jacket for easy recognition?*
- *What are some ways of communicating that you use?*

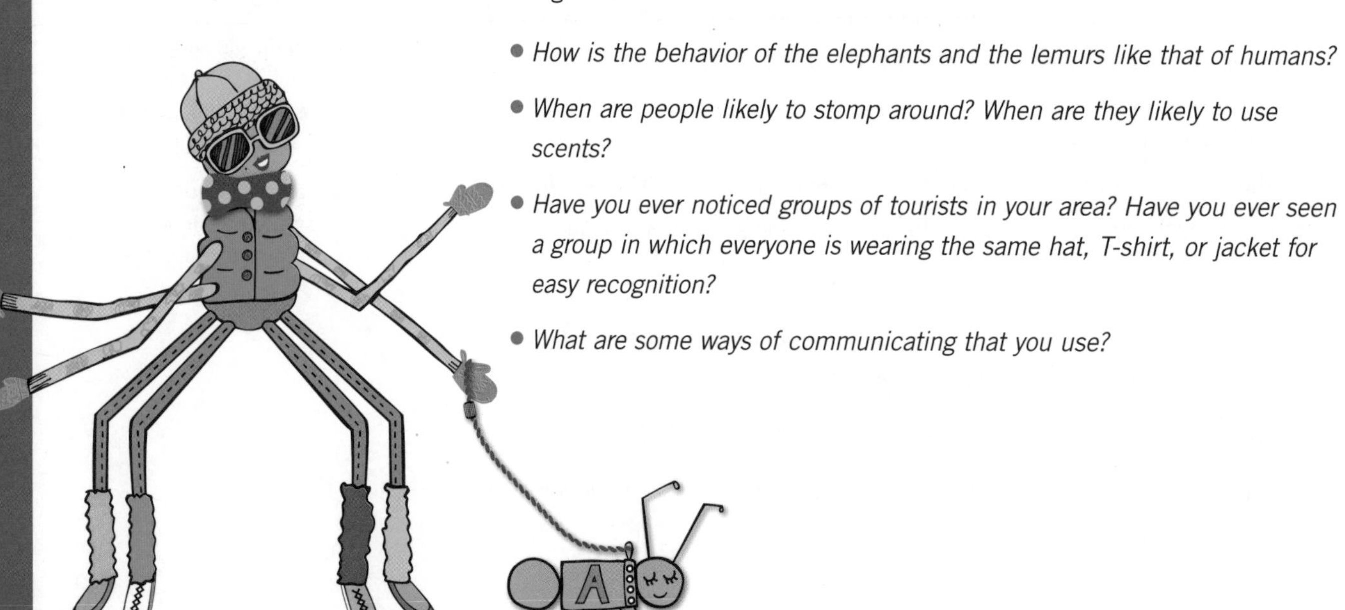

- *Name some communication strategies that help you in your relationships—or that help you handle conflicts. Do you have special conflict resolution strategies?*
- *What are some new ways of communicating that you can practice? Which ones can we commit to practicing as a team?*

Use this list as a template to get the girls creating their own list of Communication Do's and Don'ts. They'll revisit it again in Sessions 5 and 6.

DO	DON'T
Make eye contact	Make faces!
Listen	Just keep talking
Explain, calmly, why something matters to you	Yell or stomp

Closing Ceremony

Gather the girls in a circle and ask them to think about their Energy Pledges and how their wise use of energy might affect the natural world. Ask:

- *How does trying something new—like your Energy Pledge—help you grow? Is it a positive risk? In what ways?*
- *What have you learned about yourself as you've followed your pledge?*

Invite the girls to recommit to their pledges by thinking of an animal noise—a roar, a whinny, a meow, etc.—and at the count of three, bellowing out all their sounds together as a team.

Looking Ahead to Session 4

If the girls want to do the optional activity in Session 4 of making drink cozies from recycled denim, this might be a good time to tap into the Junior Friends and Family Network for help providing materials and assistance with prep work. Someone who's crafty and/or someone who has a sewing machine and knows how to use it will be a big help.

Sample Session 4

Investigating Buildings

AT A GLANCE

Goal: Juniors begin to explore energy use in buildings as they delve deeper into the science of energy and get ready to conduct an energy audit of a community building.

This session makes use of the "Investigate Your Spaces" section of the girls' book, which features many ways that people can improve the energy efficiency of buildings.

- **Opening Ceremony: Energy Pledges**
- **Plants and Energy Update**
- **Shining a Light on Lightbulbs**
- **Brrr, That's Cold! Checking for Drafts**
- **Bundle Up for Comfort: The Value of Insulation**
- **Communicating for Action**
- **Optional Activity: Hot Drink Cozies from Old Blue Jeans**
- **Closing Ceremony**

MATERIALS

- **Opening Ceremony:** Energize Awards (optional).
- **Shining a Light:** 60-watt incandescent lightbulb, 14-watt compact fluorescent bulb, 100-watt incandescent bulb, and 27-watt compact fluorescent bulb; lamp without shade; indoor/outdoor or meat thermometer.
- **Checking for Drafts:** pencils, tape, rulers; quart- or gallon-size ziplock bag, cut apart.
- **Brrr, That's Cold!:** Five clean, empty glass jars with lids, all the same size; hot water; thermometer; variety of insulating materials: wool socks, cotton T-shirt, paper, cardboard, and other items suggested by the girls.
- **Hot drink cozies (optional):** Cardboard coffee sleeves to use as a pattern (find ones with slits that fit together); scrap fabric from old blue jeans; pencil for tracing; sharp scissors; Velcro; sewing machine or needle and thread; rickrack or fabric ribbon (optional).

PREPARE AHEAD

Chat with any assistants about who will do what before and during the session.

Invite girl volunteers to assist in setting up the materials for the Shining a Light on Lightbulbs, Brrr, That's Cold, and Bundle Up for Comfort activities. Wrap the jars with whatever insulation materials you've chosen to use, leaving one unwrapped.

If doing the optional Hot Drink Cozies activity, ask girls to organize and set out the materials.

You'll find it useful to peek ahead to Sessions 5 & 6, so you can guide girls to start preparing for them at the end of this gathering.

Opening Ceremony: Energy Pledges

Gather the girls in a circle and invite them to each say one thing about how they are doing on their Energy Pledges. You might ask: *How much have you managed to reduce your energy use? Have you been able to use fewer lights? Do you think you could go even further?*

You might also ask: *If you could now add one more recycled paper bead to your necklace or bracelet of energy pledges, what would it say?*

Invite the girls to close the ceremony by recommitting as a group to making changes in their lives that save energy. Ask: *What more can we do at our team gatherings to save energy?* Remind them that saving energy at their gatherings is one step toward the Energize Award. Depending on the girls' progress, this ceremony may even be the time for them to earn the award! Ask the girls to check page 106 of their book to see if they've earned it. If so, how would they like to award one another? Perhaps each girl might want to complete this statement: *"I am most proud of how I . . . "*

WHAT'S A WATT?

A "watt" is a measurement of the rate at which electricity is used by an appliance. Watt is abbreviated simply as W. A 14-watt compact fluorescent bulb produces light comparable to a 60-watt incandescent bulb, and uses about 1/4 of the electricity. A 27-watt compact fluorescent bulb is comparable to a 100-watt incandescent bulb.

Or they might like to make an energy web in honor of all the progress Dez has made—and all the energy they've saved along the journey. As the girls form a circle, give one girl a ball of yarn. Ask her to say one thing she did to save energy, and then toss the yarn to another girl across the circle, who says something she did. Once the web of yarn connects them all, invite them to make it even stronger by continuing to pass the ball of yarn as they each say one energy-saving thing they promise to keep doing after the journey ends. As each girl states her promise, she tosses the yarn to another girl and so on around the circle. When they're all through, they can "recycle" their energy by rewinding the ball of yarn!

Plants and Energy Update

Gather the plants together, and remove the bag(s) from the test plant(s). Ask the Juniors if they can tell which plant is the control plant and which are the test plant(s). Ask: *Which one grew the best? Why do you think that is?*

Ask the girls to take out their first sketches of the plants and their leaves, and compare them to how the plants look now. Ask: *Do the test plants have more brown or discolored leaves than the control plant? What you can conclude about the importance of light to a plant's growth?*

If the team wants to continue with the experiment, put the bag(s) back on the test plant(s) and return all the plants to their resting spots until the next gathering. Encourage those taking care of the plants to keep watering them!

Shining a Light on Lightbulbs

IF YOUR GROUP IS LARGE

Set up a few stations so all the girls have a chance to get in on the action. Or include several lamps with different bulbs so girls can rotate through the activity.

Paying attention to lighting is a simple and cost-effective way to make homes, schools, and other buildings more energy-efficient. This activity emphasizes how some types of lightbulbs are more energy-efficient than others because they use less energy to produce the same amount of light.

First, engage the girls in a discussion about light in their lives. You might start the conversation with questions like:

- *Where do you see it, use it, feel it? Not just human-made light but also natural light.*
- *How and where do you enjoy light?*
- *How does the warmth of the sun make you feel when you're outdoors?*
- *How does the warmth of something baking in an oven make you feel on a cold winter day?*

Next, move the discussion to indoor lighting. You might ask:

- *How does electric light help you do what you want and need to do?*
- *Can you mentally count up the number of lights in your home? Are you surprised by the number? Does it seem like a lot?*
- *Have you ever been somewhere where the power, and all the lights, went off? What did you do?*

Then explain the next activity by saying, *Today we're going to explore lightbulbs—the light they give off and the energy they use. Paying attention to lighting is a simple and cost-effective way to make buildings more energy-efficient.*

Show the girls the various lightbulbs. Discuss how they work and how they are generally used, using the descriptions below. Be sure to stress "safety first"!

HANDLE WITH CARE!

CFLs contain mercury, considered a hazardous substance. While the amount of mercury contained in one CFL (about 5 mg) is not dangerous to humans, it's important to handle a broken mercury bulb carefully and dispose of used bulbs at a toxic waste dump and NOT with the trash.

TYPES OF LIGHTBULBS

Incandescent bulb: An electric current passes through a thin filament, heating it until it produces light. Only 10 percent of the energy used by an incandescent bulb produces light; the rest is given off as heat. **Uses:** lamps, other lighting fixtures, home appliances.	**CFL (Compact fluorescent bulb):** This gas-discharging bulb uses electricity to "excite" mercury vapors that indirectly produce visible light. CFLs are more expensive than incandescent bulbs, but last longer and can use as little as one-tenth the energy. **Uses:** Generally in lamps to replace incandescent bulbs.	**LED (Light emitting diodes):** Generates light when electric current is passed through positive and negative materials. Cool to touch, uses 2 to 10 watts of electricity, and lasts up to 30,000 hours. **Uses:** In traffic lights, in the ball that drops on New Year's Eve in Times Square, in electronic devices (cell phones, DVD players).

Ask for girls to conduct the experiment, following the "What's My Wattage" steps (page 52). One Junior might read the instructions to the group while another records the results. After all temperatures are recorded, ask:

- *Which bulb produced more heat?*
- *Is the hottest bulb also the brightest bulb?*
- *Which lightbulb is most energy-efficient, and why?* You might explain that when a device is meant to be used for lighting, not heating, any heat given off is wasted energy. So if a bulb is hot, it is not energy-efficient. You might add: *If you find that the bulbs in your home are high-energy users, you might talk with your family about switching them for ones that are more energy-efficient. That's one way to save energy!*
- *What does it feel like to now have this expert knowledge about lightbulbs? How can you use your knowledge?*

READ THAT METER!

If it is possible to view the electricity meter at your meeting place, bulbs can be switched in and out by you or another adult volunteer while the girls watch the meter. The dials will show definite differences between the types of bulbs.

What's My Wattage?

Use the chart at the bottom of this page to record your answers as you follow these steps:

1. Put a 60-watt incandescent lightbulb in the lamp and turn it on. Wait for one minute. On a scale of 1 to 10, with 10 being the brightest, how much light is produced by this bulb?

2. Hold a thermometer 6 inches above the bulb for one minute. Record the temperature. Then turn off the lamp.

3. After the bulb has cooled, remove it from the lamp and insert a 14-watt compact fluorescent lightbulb. Turn the lamp on and wait for one minute. On the same scale of 1 to 10, how much light is produced by this bulb?

4. Repeat step 2.

5. After the bulb has cooled, remove it from the lamp and insert a 100-watt lightbulb. Turn the lamp on and wait for one minute. On the same scale of 1 to 10, how much light is produced by this bulb?

6. Repeat the steps using other bulbs.

Now answer these questions:

Which bulb produced more heat?

Is the hottest bulb also the brightest bulb?

Which bulb is most energy-efficient, and why?

How does it feel to now have this expert knowledge about lightbulbs?

How can you use your knowledge?

Type of Lightbulb	Temperature	Brightness
60-watt incandescent		
14-watt compact fluorescent		
100-watt incandescent		
27-watt compact fluorescent		

Brrr, That's Cold: Checking for Drafts

Start by asking: *Do you know what a draft is?* Depending on the girls' answers, you might explain: *A draft is air moving into or out of a building in a place where it's not wanted, such as around a window, door, or electrical outlet. The places where cables enter buildings can also be drafty.*

If a building has drafty places, energy is being wasted because it's escaping from doors or windows. It means outdoor air is coming inside and indoor air is going out. So you may be heating or cooling that unwanted outdoor air, and your heated or cooled indoor air is escaping. This means that rooms can lose heat in winter or lose their cooled air in summer. All of this wastes energy.

Then you might say: OK, now, let's *Get Moving!* and check for drafts.

- If you have a large group, divide the girls into teams.
- Then pass out the draft tool materials (pencils, plastic wrap or baggies, and tape) and instruct the teams to cut a 6-inch x 12-inch piece of plastic wrap (or plastic baggie), and tape one short edge of the plastic to a pencil, letting the rest hang free. The girls can gently blow on the plastic wrap to see how sensitive it is to the flow of air.
- Encourage the girls to move about the room looking for places that may feel drafty. You might ask: *Where would you check for drafts? Windows? Doors? Anywhere else?* (Electrical outlets would be another place.)

Ask: *Do you see cracks in the walls, window air-conditioning units, or other openings to the outside? For each area, use your pencil draft tool to check for airflow. Record your observations in your chart (photocopied from page 56). Rate each drafty area as either "no draft" or small, medium, or large.*

NOW, WHAT CAN YOU DO?

If the girls found drafty areas, invite them to brainstorm how to improve them. Could they design something new and unique to fix them? Encourage them to brainstorm and test out their ideas at home. Then, perhaps they can speak to the building manager to present their ideas. Ask:

- *Now that you know about this particular building, where else might you want to investigate improving energy use?* Encourage the girls to think about a building in their community that they might like to visit to perform an energy audit. Ask:
- *How can your investigating lead to lasting change? Why is that important?*
- *How does creating lasting change allow you to live the Girl Scout Law?*
- *When you learn something new, why is it important to pass your new knowledge along to others?*

INVESTIGATE!

You might introduce this activity by discussing what it means to investigate: *"Investigate" means to examine, to study, to check out the facts. Fictional detectives like Nancy Drew in the mystery series and Adrian Monk, on the TV show "Monk," investigate, and so can you!*

You don't have to investigate crimes or a mysteries. If something makes you curious or you're puzzled about something, you can investigate it. That's what leaders do!

On this journey, investigating ways to make buildings more energy-efficient is also a step toward earning the Investigate award!

BE PREPARED

The Juniors' team audit of a community building will have them investigating even more about energy.

Take a moment after this "Now, What Can You Do?" discussion for the girls to flip through the "Be Prepared for Your Energy Audit" sections of their book (starting on page 70). These will help them "Be Prepared" for the audit day.

Ask: *What might you investigate at home before our big energy audit? What changes might you ask your family about? Maybe you can complete the first two steps to your Investigate Award!*

Say: *When you speak out about something important, you are being an "advocate." Check out the "Relate and Communicate!" tips that we came up with (Session 3) and have been practicing. What new communication tips might be important when we talk to people about using energy wisely?* Invite girls to add new tips to their list.

Bundle Up for Comfort: The Value of Insulation

- Insulation improves the use of energy in a building because it keeps heat inside in the winter and air-conditioning inside in the summer.
- Guide the girls through an examination of the various insulation materials that have been wrapped around the jars. Ask: Which insulating material do you think will work best? Ask them to record their predictions.
- Next, fill all the jars with hot water. Record the starting temperature, then seal the jars with lids.
- Set the unwrapped jar aside. This is your "control" jar.
- Let the jars sit for 30 minutes.
- Remove the lids and record the temperature of the water in each jar.

REFLECTION AND BRAINSTORM

Now ask the girls: *Based on the results you found, what type of insulation do you think works best?* You might also ask: *Do you have ideas for designing a new or better type of insulation?* You might mention that recycled denim—old blue jeans!—is now being used as insulation for homes and other buildings. Point out that it is used in the California Academy of Sciences, which is described and pictured on page 64 of their book.

KEEPING WARM

Keep warm by wearing clothes that hold in the warmth of the body. Puffy, woolly materials do this because they have air in their thickness. Air is like a wall. It holds the heat of your body in and keeps the cold air out. Many layers of clothes are warmer than one thick layer, because there are layers of air between each.

Try this: Fill two glasses with water. On the water in one glass, lay a small piece of wool cloth. On the water in the other glass, lay a small piece of cotton cloth. Look for little bubbles around the wool. These bubbles are air that was in the puffy wool. The air in the wool keeps it afloat longer.

MAKE IT PRACTICAL

This activity will also work well if the girls are looking at ways to keep themselves warm or their lunches cold. So an alternative to the assorted materials could be their own lunch boxes and water bottles. Have the girls brainstorm on ways to test how well these items insulate things. Can they place them in the refrigerator? Or outside if this is done in the winter or warmer months?

FROM THE ARCHIVES

Depending on the team's time and interest, invite the girls to test out the "Keeping Warm" experiments at right, which are adapted from the *Junior Girl Scout Handbook* of 1963, which demonstrate the value of insulation. You might bring the experiments home (back to energy efficiency in buildings) by asking: *Do you ever wear a sweater indoors rather than turning up the heat?*

On a camping trip, change to dry clothes when you go to bed because your day clothes will be damp from perspiration—even if you were not hot. When the moisture from your day clothes evaporates, it makes you cold.

Experiment with your clothes to see which combinations are warmest in windy weather. You can keep the cold wind away from your body by wearing jackets and other clothing that block the wind. When it's windy, a light jacket worn over two sweaters may be warmer than a heavy coat worn with no sweaters.

Crafty Option: Hot Drink Cozies from Old Blue Jeans

Hot drink cozies are a form of insulation—and they conserve resources by replacing paper cozies that usually get thrown away after each use. Encourage the girls to gift the cozies they make to an adult who enjoys coffee or tea. Encourage the girls to think about how nice it is to be able to give away something they've made, especially something that saves resources. Ask: *Do you think this creates a lasting change—by cutting back on the use of paper cup holders—that might catch on in a bigger way?*

DANGER! DENIM!

Denim is a thick fabric and hard to cut, especially through the seams. Caution the girls to take care when cutting it and offer any needed assistance.

To make the cozies: Invite the girls to trace a standard cardboard cup sleeve onto the denim to create a "pattern" they can cut out. Then they can sew one piece of Velcro on one end of the denim and a second piece of Velcro on the inside of the denim at the other end, making sure that the two pieces overlap and fit securely around a cup. If they like, the girls can put a decorative row of stitches around the edge of the fabric (about a quarter- to a half-inch), sew on rickrack or fabric ribbon, or decorate with buttons and/or embroidery.

Closing Ceremony

Invite the girls to form a circle and name one thing about energy that they are curious about and would like to investigate. Ask: *How might you take your curiosity and turn it into some lasting change?* (For an example, you might say: *I wonder how much energy my family uses to heat our home in winter. I'll look at my gas bill with my family and talk about ways to cut down that would save us money every month.*)

End the ceremony by asking: *By following your curiosity, you may end up learning something new. How does it feel to try to learn something new? Being curious is good because ______________________________ ?*

CHECKING FOR DRAFTS				
LOCATION	NO DRAFT	SMALL DRAFT	MEDIUM DRAFT	LARGE DRAFT

What ideas do you have for stopping the drafts you found?

INSULATOR	STARTING TEMPERATURE	ENDING TEMPERATURE	DIFFERENCE IN TEMPERATURE
No insulation			
Wool socks			
Cotton T-shirt			
Paper			
Cardboard			
(your suggestion)			

Sample Sessions 5 & 6

The Energy Audit

AT A GLANCE

Goal: The Juniors conduct an energy audit of a community building in order to educate and inspire on the importance of energy efficiency.

- **Energy Audit**
- **Communication Maze**
- **Relate and Communicate: Favorite Tips**
- **Communicate with Style (and Energetically)**
- **Say It Like an Advocate**
- **Letter Writing Do's and Don'ts**

Now that girls have dipped into the energy use in buildings, they're ready to investigate further!

Buildings are responsible for about 40 percent of all energy use, so making buildings more energy-efficient is important.

The activities outlined here can take place over Sessions 5 and 6, and even spill over into Session 7, depending on the Juniors' time and interest.

MATERIALS

- **Energy Audit:** Copies of "The Big Questions"; computer with Internet connection, if available.
- Copies of any other activity sheets the Juniors will be enjoying (see pages 62–75).

Why an Energy Audit?

Conducting an energy audit of a community building (even the place the Junior team meets) will allow girls to:

- Meet new people and learn about all the kinds of work involved in creating and running energy-efficient buildings (that's important!)
- Use their leadership skills and values to make buildings more efficient (that's good for Earth)
- Try something new (that helps girls embrace challenges)
- Share their new expertise with others! (that boosts confidence)
- Advocate for change! (that's good for people and Earth)

SEEK EXPERT HELP!

Ask your Girl Scout council whether experts are available in your area to work directly with the girls during their Energy Audit.

MORE JUNIORS, MORE ENERGY!

Consider networking with other Girl Scout Junior teams who are journeying through *GET MOVING!* You might link up to investigate a building together.

What the Juniors Will Do

Over the next few sessions (however many you and the girls plan to spend on the Energy Audit), the team will:

- Identify a building to audit and a person who knows the building well enough to show the team around and help it gather the information (including energy bills) it will need for the audit.
- Practice good communication skills—so the girls can obtain information that interests them and talk about what they have learned.
- Conduct an audit in the chosen building, using the list of questions provided in this guide.
- Analyze the building's impact on the environment and use what they learn to offer ideas for improving the building's energy efficiency.
- Share the experience with the building's manager and public officials, so that even more buildings can become energy-efficient!

The team will really be "on the move" now, so this next section of your book is set up to guide you on what to do, and then offer you flexibility as to how your team goes about it.

Get to It! The Building Audit: Step by Step

Don't worry! You don't need to be a building management expert. You can investigate right along with girls. Here's what you need to do:

BEFORE VISITING THE BUILDING

- **Make and confirm arrangements** for the team to visit a community building. It can be where you meet, a library, school, local business, or somewhere else you know. Use the Friends and Family Network to find the best building possible. Someone will know someone who can assist in making the arrangements! You'll want to contact the building manager, who will know the building's inner workings (heating, electricity, etc.) and can show the team around. Plan enough time for a thorough tour. Explain that you are a Girl Scout volunteer guiding girls to learn more about how to use energy wisely.
- **Be safe! Tap into your Friends and Family Network** for other adults to join you and the team. Follow your council's safety guidelines for outings.
- **Give the building manager the audit questions** (pages 71–72) ahead of time. That way they can have as much information as possible ready when you visit. But don't worry: If the girls can't get answers to all the questions, they'll still learn a lot from the answers they do find!

- **Depending on who you identify** to guide the team, the energy audit may also be an opportunity for girls to explore the many career options that relate to energy. Notice the "Hello I'm a . . ." activity sheet on page 68. Give it to the building manager and other experts joining you, and also to the girls. That way, the girls can ask questions and the experts can be ready with some answers.

- **If your building investigation** takes you and the team on a field visit (someplace other than where you usually meet), take time to do something else the girls might also enjoy. You might check out the energy people are using. You might venture outdoors, try a new food, or make like Dez and check out any vintage shops in the area to get a firsthand look at fashionable recycling.

- **Talk to girls and their families** about who can take charge of recording the information the girls collect during the energy audit. A girl/adult pair or two would be ideal. If you have immediate access to a computer with an Internet connection, the whole team can take turns entering the data. If not, the volunteer pair will write all the needed data and then enter it online on http://energy.trane.com. They can then share the results with the team so that all the girls can decide what can be improved.

- **Boost communication confidence** by engaging the girls in the "Communicate with Style (and Energetically!)" activity on page 63–64. Or do this after the audit, in preparation for when the girls share what they have learned.

WHILE VISITING THE BUILDING

- **Encourage the girls to be curious** and to use what they've learned all along the way (lights, drafts, insulation, etc.). Encourage them to look for as many clues to the building's energy use as possible. It's important for the Juniors to observe the building's energy use with their own eyes.

- **Coach girls to find answers** to as many of the audit questions as possible.

- **Make the audit an active experience.** Depending on your group's size and logistics, girls can take turns asking questions, and they can work in small teams, too.

- **Remind your girl/adult data input team** to record all the audit information, so the team can learn how the building stacks up.

AFTER VISITING THE BUILDING

Have the girl/adult team enter the building's energy information online at http://energy.trane.com and then see what girls can learn by analyzing the results. What ideas can they come up with for making the building more energy-efficient?

As the girls look over their data, encourage them to think about their own observations, and their ideas about energy efficiency, now that they're gaining expert knowledge. Ask:

- *What did you learn? How does it feel to do something totally different?*
- *Who did you meet? What was most interesting about the experience?*

If you have expert advice, tap it. Otherwise use what you and the girls can analyze and brainstorm on your own.

SHARING THE INSIGHTS

Once the girls have ideas about how the building's energy use can be improved, they can report back to the building manager or other building officials. Can the building manager visit your team meeting? Or will the team make another visit to the building to share its report? If logistics make a meeting impossible, the girls can send an e-mail to the building official(s) with an outline of their recommendations for improving energy use.

Encourage the girls to use the best information they captured to report back to the building manager. Be sure to ask:

- *What else might you like to ask and learn?*
- *How does it feel to speak up and share your ideas?*

Girls can share their knowledge further by writing to local officials and legislators. The letter-writing tips on pages 74–75 will give them all the support they need.

GETTING ENERGIZED FOR THE AUDIT!

The following pages hold your Audit Tool Kit—all the activities and handouts you and the team will need. Mix and match the order of how you use them over one, two, or more sessions, based on your team's plans and logistics and the girls' interest. (The Tool Kit is also available online at in the Journeys section of girlscouts.org for easy photocopying.)

And don't forget the reward! Girls will have earned the Investigate Award by following the steps on page 107 of their book! Have a great ceremony where girls can celebrate their newfound expertise!

The Audit Tool Kit: What's In It

- **Communicate with Style,** pages 63–64: Use these tips to build the girls' confidence about speaking up—whether to ask a question, describe what the team is doing, or report back on energy use.
- **Say it Like an Advocate,** page 65: These "Do's and Don'ts" offer more great communication tips for the girls.

- **Communication Maze,** page 66–67: This fun activity will get girls thinking even more about the importance of clear communication.
- **Hello I'm an Engineer (or a ...),** **How Many Engineers Do You Know, and Asking Good Questions, Being a Good Partner,** pages 68–69: Use all of these to encourage girls to explore interesting career possibilities they may never have thought of!
- **The Big Questions,** pages 70–71: All the questions the girls need to investigate a building's energy use and come to some conclusions about improving its energy efficiency! **Remember:** You don't need expertise in energy systems to guide the girls through the audit. Just follow the fill-in-the-blank questions during the audit.
- **How Does Your Building Stack Up?**, page 72: These questions and suggestions will help the girls analyze the energy data from their building.
- **Advocate for Change!**, page 73: These questions and suggestions will help the girls organize what they've learned into energy-saving recommendations for the building's staff.

If it helps, use this handy sheet to "check off" all your audit steps:

OUR BUILDING AUDIT PLAN

- [ ] Building we'll visit ______________________________
- [ ] Date/time ______________________________
- [ ] Building manager______________________________
- [ ] Other potential energy expert(s) ______________________________

____

- [ ] Copies of audit materials ______________________________
- [ ] Girl/volunteer team to record and input audit data online ______________

____

- [ ] How/when we'll give ideas to the building ______________________________

____

____

____

ENERGY AUDIT: THE BIG QUESTIONS

Encourage the girls to give the building manager the list of questions ahead of the audit so she can gather all the information needed before the Juniors visit.

Some of the questions are also in "My Building Bio" on page 74 of the girls' book, so the Juniors can keep a permanent record of the building they audited.

For the questions about utility bills, monthly or annual bills can be used.

Be sure to secure time with the staff for a walk-through of the building with your Junior team. That way the girls can see firsthand how and where energy is in use.

TIPS FOR TALKING

As the girls prepare to investigate the building's energy use, and then share what they're learning in order to educate and inspire others, they'll be networking with building managers and other community experts. Use the "Communicate with Style (and Energetically)" tips, starting on page 63, which are meant to be photocopied and given to the girls, to get the Juniors putting their best feet forward.

Communication Maze

The Juniors can enjoy this activity along with their energy investigations, or at any time. They might even want to include it in their final celebration.

Start by emphasizing that communicating clearly is an important skill for engineers, and for leaders. You might say: *Team members in all walks of life need to understand directions they are given. Otherwise they may go off in the wrong direction, and that's not energy-efficient. This is especially important in engineering.*

Then say: *Now, let's pair up and test our communication skills.*

Give one girl in each pair a pencil and a copy of the first maze (page 66) face down. Then explain the game:

- *The girl with the copy of the maze will follow directions from her partner. When it's time, she'll close her eyes and flip over the maze paper.*
- *Her partner will place the pencil point at the start of the maze and talk her through it. The partner will give directions like, "Start drawing a line toward you. Now stop. Now go left. Further, further, stop. Go left . . . " until the girl with her eyes closed reaches the end of the maze.*
- *When the first girl has finished, the partners switch roles and try the second maze (page 67).*

When they're finished, engage the girls in a discussion about the game and communication. Ask questions like:

- *Was it harder or easier than you thought it would be? Why?*
- *If things didn't go as smoothly as you would have liked, what would you change about your directions next time? What would you want your partner to do differently?*
- *How can you use your energy to communicate better?*

Relate and Communicate: Favorite Tips

Remember the Do's and Don'ts the girls started in Session 3? What else might they add now that they've been talking to more people and sharing their energy expertise? Ask:

- *What Relate and Communicate Tips have you been practicing with your Junior Team?*
- *What Relate and Communicate Tips do you use at school?*
- *What is your top Relate and Communicate Tip that you want all your friends to use, too? How does it help?*

Communicate with Style (and Energetically!)

As you Energize, Investigate, and Innovate, you'll find that you're networking with a lot of people. You'll talk to people who manage buildings. You'll share your innovative ideas to get others saving energy. You may even interview people to find out what they know—so you can know it, too!

You can communicate in all kinds of ways—by e-mailing or writing letters that you mail at the post office. You might make a presentation or create an exhibit.

No matter what you do, here are five tips to make sure you get the maximum impact when you deliver your message! Once you master these, maybe you can add a tip or two of your own!

1. Say Who You Are

You are a Girl Scout! You are on a leadership journey and you are learning to use energy wisely and inspiring others to save energy, too.

Who else are you? You are part of a major force for improving the environment! There are 2.6 million Girl Scouts. About 500,000 of them are your age, and you are all trying to do something good for Earth. That's powerful. So say it loud! Say it proud!

2. Say What You Need

Be clear and specific about what you want others to know or do. If you ask, "Can you help us save energy?" it might sound very big and very hard. But if you say, "Can you help us start a once-a-week walking club for our school?" you are breaking it down and giving people a specific idea that will feel doable. You can negotiate from there! Just remember: Vague is not in vogue, and specificity is in style!

3. Show What You Know!

If the building management shuts the lights off a little earlier or stops a draft, how will that help save energy? If families carpool to the Girl Scout meeting or the library, how will that help the Earth? If your school stops wasting paper, what will be the impact?

You don't need to say a lot. Just make what you say really carry some oomph. Use the facts you are learning along this journey—the ones that mean the most to you—to inspire others. Imagine you are the anchorperson on a news program. What are the few most important things you need to say to get people to stop and listen? How can you describe how your idea for improving energy use can help people—and save energy, too? Practice what you want to say!

4. Let Your Curiosity Show!

Other people love to talk about what they know, so ask plenty of questions wherever you go. Just remember, the "W's" (who, what, when, where, why) are usually better then the "D's" (do, did, don't) question openers. Why? Check this out:

- Do you manage the building's heating system?
- Did you ever set up a bike club?

Notice that if you start with these "D's," the likely response will be "yes" or "no" and then you might get stuck! There's nowhere to go!

But if you start with a W, you'll find yourself in a richer conversation.

- What kind of heating system is best for this building?
- Why would a biking club be good in this town?

5. Be Kind and Positive

OK. Face it. If someone tells you, "Hey, you lazy thing! You better get walking," you are probably not going to be very happy about it, right? And, you won't feel very motivated to Get Moving!

But if someone invites you like this: "Guess what? Great news! Every Friday afternoon we are going to have a walking club and you can be a member," you might stick around to hear some more, right?

No one wants to feel put down, but everyone loves a positive boost. So when you suggest that a building manager (or your mom!) repair a drafty space, don't say: "Ah, duh . . . we're freezing in here! What are you going to do about it?" Do say: "I'd really love to help save energy—and money—here. I noticed that it's very drafty in this spot, and I'll bet we could think of a way to fix it. Want to try?"

Say It Like an Advocate!

You've heard of "fashion don'ts"? Well, below are some communication don'ts. Think them through and replace each with a "communication do." Then add your own communication tip (or two) to the list and share them with friends.

1. You want your school to sponsor some walking days. You meet with the principal, and you start by saying, "We are in a group and we have an idea and maybe you could help?"

(Hint: Who is we? What is your idea? Why is it important?)

2. You are asking everyone in your class to stop wasting paper. So you say, "Look at all the paper in the trash bin! This is really bad! Haven't you ever heard of *re-use*?"

(Hint: Be kind! Be positive!)

3. You are interviewing a building manager to find out how the building could save energy. You ask: "Did you replace the heating system recently?" She says, "Yes." Now you are stuck because you don't know if that helped or not. Start over with a better question.

(Hint: Try a W or an H question!)

MAKE YOUR OWN DON'TS—AND SOLVE THEM, TOO!

Now that you're a successful communicator, what are your top three do's?

1.

2.

3.

Relate and Communicate: Favorite Tips

1.

2.

3.

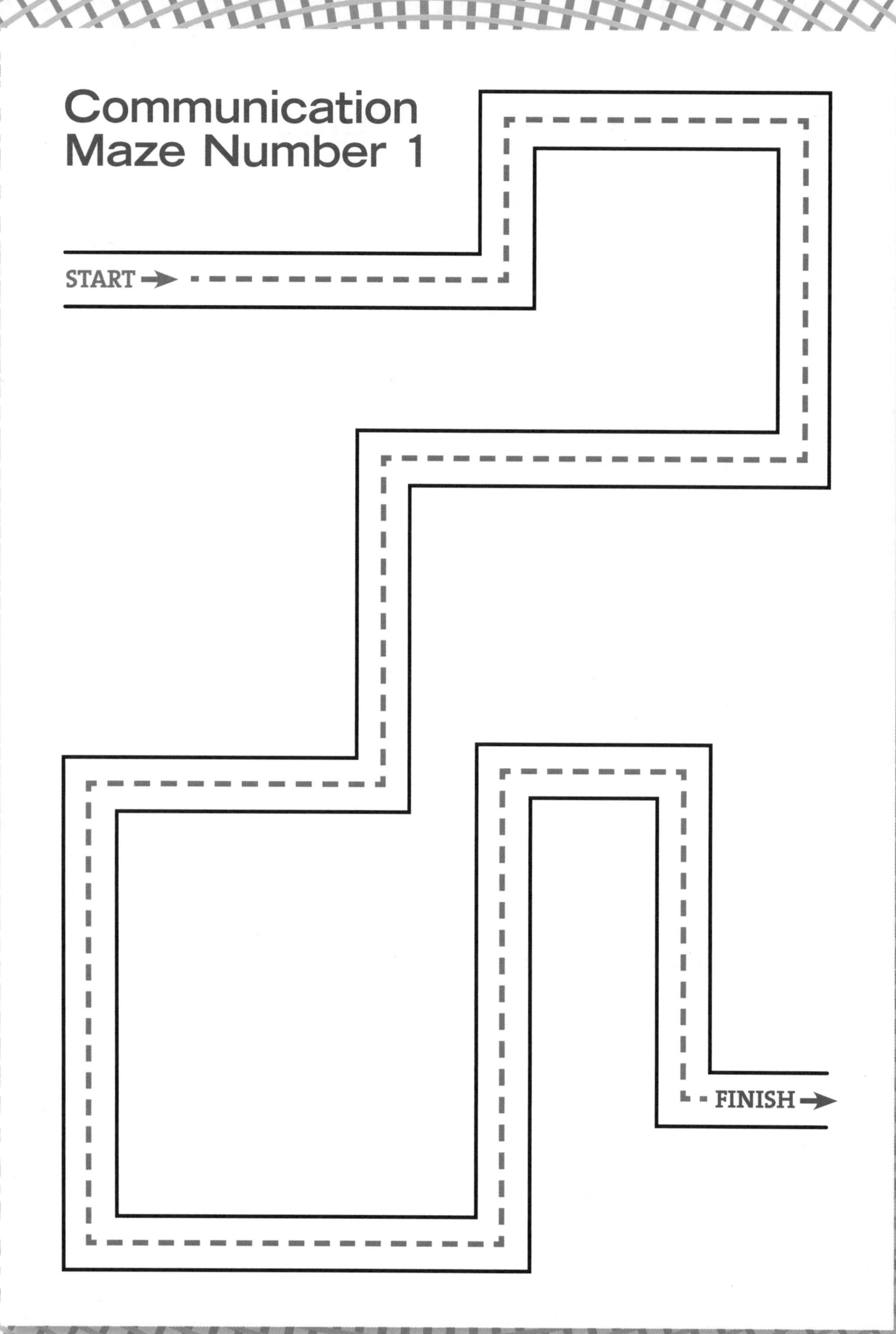
Communication
Maze Number 1
START
FINISH

Communication Maze Number 2

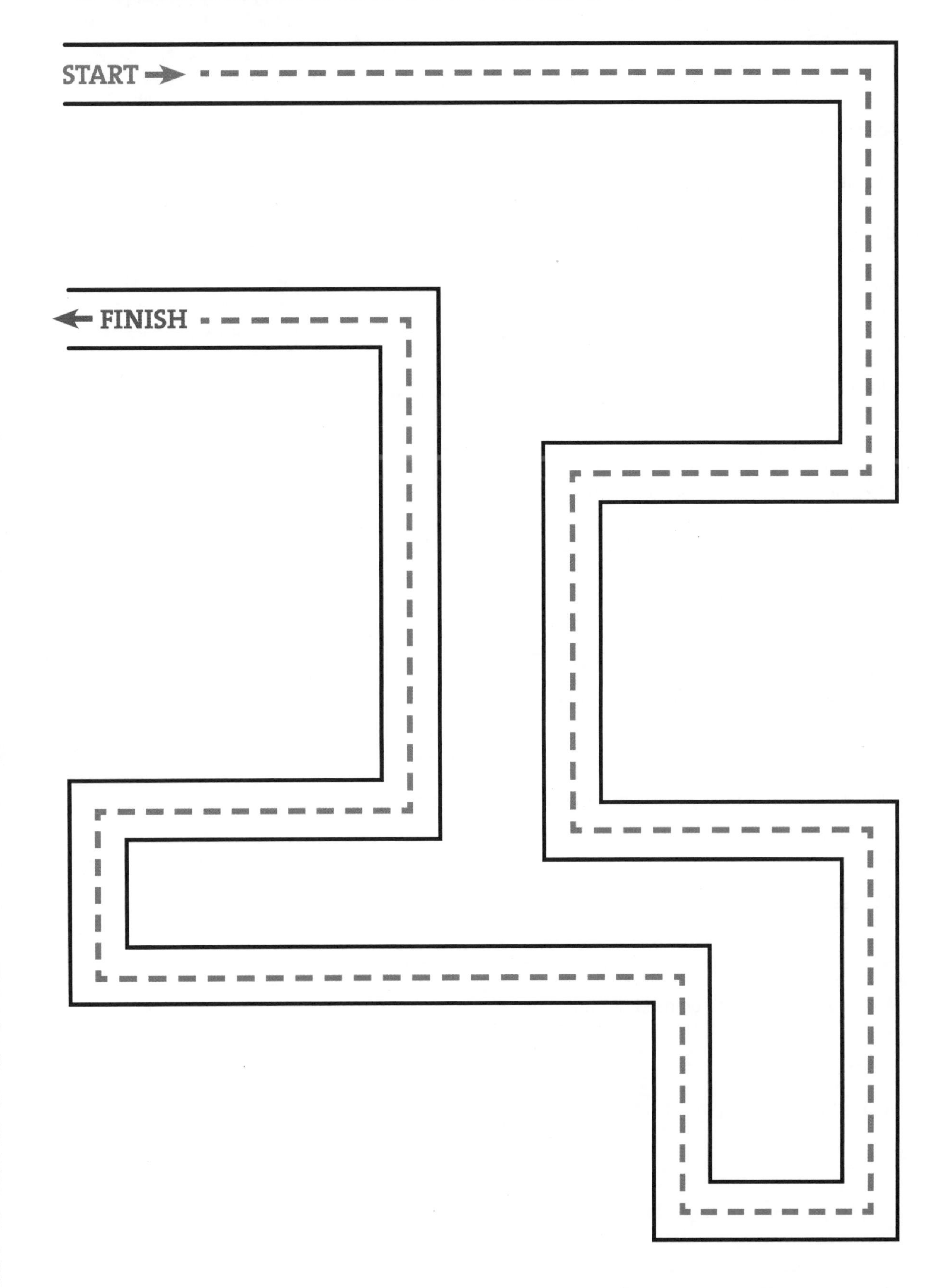

Hello! I'm an Engineer! (or a ______________________)

Name: ______________________

Job Title: ______________________

Company: ______________________

Years in current job: ______________________

BUILDING ENERGY EXPERTS!

Create your own "profile" to share with the girls as you guide them through their energy audit.

GIRL SCOUT JUNIORS!

Be curious! What would you like to know about careers involving energy? Now's the time to ask!

When did you decide that you wanted to become an engineer?

____

What kind of education and training does an engineer have to have?

____

What's a typical workday like for you?

____

What's your favorite thing about your work?

____

What's the weirdest thing that ever happened to you on the job?

____

What saps your energy on the job?

____

What would people be surprised to learn about you?

____

What's the most energizing thing you've done on the job?

____

What advice would you give to girls who want to be engineers?

____

How Many Engineers Do You Know?

Engineers use science and math to find answers to problems in everyday life. They invent, design, and develop new products. They also help fix things and keep them running. Match the kind of engineer with what she might invent, improve, or keep running.

Aerospace engineer	paints, fuels, fibers, clean technology
Agricultural and biological engineer	highways, skyscrapers, bridges
Biomedical engineer	energy-efficient heating and cooling systems for buildings
Chemical engineer	medical instruments, artificial organs
Civil engineer	anything involving a mechanical process, from a hybrid car engine to a child's toy
Electrical engineer	aircraft, satellite, rockets
Energy engineer	plants, fertilizers, sprinkler systems
Mechanical engineer	power grids, MP3 players, roller coasters

Asking Good Questions, Being a Good Partner

When you're wondering how to make a building more energy-efficient, what questions need to be asked? What do you investigate? How can you innovate? Working with a sales engineer can help.

Tiffany Zimmer earned her college degree in mechanical engineering. She now helps people choose the correct heating and cooling systems for their buildings.

"I've always been interested in how things work," Tiffany says. "I also wanted to find a field where I felt like I was helping people."

Tiffany's job title is sales engineer. She helps customers select and purchase the heating, ventilation, and air conditioning (HVAC) systems that best fit the needs of their office buildings, or the hospitals, universities, and retail stores they manage. "The most rewarding thing about my work is when my customers view me as a partner and as a resource," she says.

Tiffany's father was a mechanical engineer who worked in the oil industry, so her family moved around a lot. She lived in Colorado, Alaska, Illinois, Wyoming, and Cairo, Egypt. She credits the moves with helping her to develop an outgoing personality and good communication skills, both of which are useful in her work.

- What kind of engineer might you like to be?
- How do you think engineers help people and the planet?
- Where would you like to try living and working someday?
- What's your best communication skill?

THE BIG QUESTIONS

What Can You Learn . . .

From a Building's HISTORY

- Name:
- Location (city and state):
- Type (commercial, school, government, etc.):
- Year built:
- Average number of people using the building each day:
- Average number of hours each week the building is used:
- Months per year the building is used:
- Square footage of the building's floor space that is heated/cooled:
- Is the building air-conditioned?
- What type of system heats and cools the building? (If the type cannot be determined, enter "constant volume," the most commonly installed system.)
- What updates have been done to make the building more energy-efficient (lighting, insulation, roofing, windows, heating/cooling systems, hot water heaters, etc.) and when?
- Besides electricity, what fuel(s) is/are used to heat/cool the building?

INTERNET ACCESS

If your team has assigned the data inputting to a girl/adult pair, be sure they do so at http://energy.trane.com and then deliver the results to the full team so all the girls can analyze it.

Don't have online access? Check with a library or a school, or a Network member who might help.

Still no luck? Simply mail the team's answered questions to Ingersoll Rand. An analysis of your energy audit will be returned to you in two to three weeks. (Mail to: Ingersoll Rand, ATTN: Greg Jakobek, 800 E. Beaty Street, Davidson, NC 28036)

From a Building's BILLS

Electricity: When reading the annual or monthly electric bill, what is the total cost? What is the kWh or $/kWh?

Natural Gas: On the natural gas bill, what is the total cost? What are the Therms or $/Therms?

Propane: What is the total cost? What are the gallons or $/gallons?

Fuel Oil: What is the total cost? What are the gallons or $/gallons?

Get answers to as many of the questions on the next page as you can. The answer may be "I don't know"; that's OK. But the more questions you can answer, the more informative your audit will be!

From a Building's EXPERTS

- Does the building have an automation system?
- How much do you lower the temperature at night or during off-peak times?
- Do you have any indoor daylighting sensors and/or occupancy controls?
- How many elevators and/or escalators are typically in operation in your facility?
- Do you have any service agreements on your HVAC or automation equipment?
- How is the facility's water heated?
- What is the average temperature setting for your hot water system?
- How do your facility's hot water needs compare to those of similar facilities? (For example, a hotel's hot water usage typically would be considered "high" if it had both in-house laundry and a swimming pool. If it had one or the other, it would be "average"; neither would be "low.")
- What type of fuel is used for commercial cooking?
- What is your overall interior lighting load?
- Looking at the outside of the building, what percentage of the wall area is windows?
- What type of windows do you have (overall window U-value)?
- What is the color or tint of the windows (average shading coefficient)?
- How are the walls constructed (overall R-value)?
- How is the roof constructed (overall R-value)?
- What is the efficiency of the heating system?
- How much of the building is mechanically cooled?
- How efficient is the cooling system?
- If the building has a chiller plant, what type of chillers are used?

TIME TO ANALYZE YOUR DATA

Now that have you've gathered all this information, what can it tell you? Plenty! But first you'll need some help analyzing it all. With your team, go online and visit:

Site: http://energy.trane.com

User name: Girl Scouts

Password: Trane

Input all of your building's information and you'll receive your building's energy ratings. The online program will tell you how well your building is doing on energy use compared to other buildings in your area and around the country. It will also tell you the environmental impact of the building. Be sure to name your building so that you can easily find your data whenever you revisit the Web site.

How Does Your Building Stack Up?

Now that you've received the analysis of your building's energy use, what can you learn from it? Look it over carefully.

- How does what you read in your analysis match up with what you observed at the building yourself? How many points made in the analysis could you have come up with from your visual audit alone?
- If you noticed leaks, are you seeing that your building is not as efficient as others in your area?
- What kind of improvements might you suggest to those who run the building?

Now, see if you can summarize the main points of the analysis well enough to present them to the building staff. You'll want to let the staff know how their building is doing compared to similar buildings in their area. And you'll want to share your recommendations on how they can make the building even more energy-efficient (there is always room for improvement!).

To get started, use the results from the Energy Analyzer you received to fill in the blanks below.

- Annual energy costs for your building are ____________________ per square foot.
- This is ___________ % __________________ [more/less] (circle one) than for a similar building in our region.
- In contrast, energy costs for a new, energy-efficient building of this type would be approximately ____________________ per square foot.

Taking a Closer Look at the Building's Energy Use

As you continue to review the energy breakdown for your building, focus on those areas that are the most costly. They offer the potential for the greatest savings. In your case, these are ____________________ and ____________________. (Insert the two most costly areas of energy use based on the results of your energy audit.)

The Building's Carbon Footprint

Based on our audit, the building produces _________ metric tons of CO_2 each year. The industry average for buildings like this is _________ metric tons of CO_2 each year; for a highly efficient building, it's __________ metric tons of CO_2 each year. A typical automobile produces just __________ metric tons of CO_2 each year.

Advocate for Change!

As you share what you've learned—to encourage others to act on your recommendations—use the best communication strategies possible to achieve your goal. These questions will help you prepare:

- In performing an audit of your building, we were surprised to learn that:

  ____

- One thing we noticed that you could do right away to start becoming more energy-efficient is:

  ____

We recommend the following energy-efficiency suggestions for you, which you can put in place as time and budget allows:

1.

2.

3.

(Be sure to end your presentation by thanking the building staff for their time in assisting you as you learned about energy efficiency. Ask if they have any questions for you.)

Turn Your Knowledge into Action!

Now it's time to share your knowledge with others—to inspire them to work for energy efficiency too.

- Write to your legislator to let her know about what you've done and why it's important. Use the sample letter and all your writing tips to get you started.
- Put your new energy knowledge to use by helping neighbors and friends conduct audits of buildings in their community. You might even create business cards for your Junior group using your recycled paper! And if you've worked with an energy expert, be sure to ask for some tips!

To play our Eco Dream House interactive game, visit ForGirls.GirlScouts.org.

Letter Writing Do's and Don'ts

1. Think about what you want to accomplish with the letter.
2. Find out the correct spelling of the name, title, and mailing address of the person you are writing to.
3. Use the proper business letter format, as in the GOOD SAMPLE LETTERS on the next page.
4. Write a draft of the letter and review it for spelling, grammar, and factual information. Have someone else review it, too.
5. If the letter is handwritten, make it neat and easy to read. If typed, check it for typos.
6. Include a way for the recipient of the letter to contact an adult volunteer from your group.
7. End the letter by thanking the person for her time, and let her know you look forward to her reply.

Address the letter to "Dear ______________," and use the person's last name and preferred courtesy title such as Ms. or Mr.

It should be "you're" not "your." Check for typos!

Be clear about what you want. Write: "Now my group would like you to visit with us to talk about energy and energy efficiency."

Hello!

How are you doing? I hope your doing fine. I am having so much fun with my Girl Scout Junior group. We made recycled paper and saved all our trash for one week. We made a pledge to walk places as much as we can.

Now my group wants you to come and visit us. It'll be fun.

We meet at the hall on Wednesday after school. So come if you can! You can call me if you want to. Thankyou for your help.

Signed,

[Your name]

"Thank you" should be two words.

Ask the person to call you to let you know whether or not she can come. That way, your Girl Scout group can "Be Prepared"!

Use "Sincerely" rather than "Signed."

Where do you want the person to be and at what time? Use the full name and the address of the place where you meet and the date and time you want her to arrive.

Include a phone number where the person can reach your trusty adult volunteer.

Two Good Sample Letters

[Your name, address, and the date, all on separate lines]

Samantha Speaker
1234 Friendly Lane
Your Town, ST 12345

Dear Ms. Speaker:

My Junior Girl Scout group is on a journey all about energy. We are interested in learning from people who know about the energy in our world, and ways to use it wisely.

We would like to invite you to visit with us to talk about your research with elephants and other endangered animals. We get together every Wednesday afternoon at the Community Hall in Your Town. We will be meeting every week through the end of June. It would be wonderful if you could join us at one of our gatherings.

Thank you for taking the time to consider the invitation. You can call, [adult's name] at [phone number] to let me know if and when you are available. I look forward to hearing from you.

Sincerely,

[Your name]

RELATE AND COMMUNICATE

When is it best to use a letter to communicate? When is electronic communication (texting, e-mail) more useful? When is it not?

How do Communication Do's differ for letters and talking? How are they similar?

[Your name, address, and the date, all on separate lines]

The Honorable [full name]
United States Senate (or House of Representatives)
Washington, DC 20515

Dear Senator [last name]:

As a Girl Scout in (insert your town name), I have become aware of how important it is for all of us to use energy wisely in everything we do.

One of the values of the Girl Scout Law is to use resources wisely. Another is to make the world a better place. I urge you to join us in living those values by making buildings—especially school buildings—more energy-efficient.

Buildings are a big part of our lives. Most people hardly notice how energy-efficient or inefficient buildings are. But I have learned that energy use in buildings has a significant impact on our planet's precious resources.

The first step in making buildings more energy-efficient is to perform an audit, as I have done to earn the Girl Scout Junior Investigate Award. Recently, my Girl Scout group has been searching for ways to make buildings, such as (give an example of the building that you examined), less wasteful.

Please ask the State Department of Education to take action by conducting energy audits of all school buildings. I'm counting on you to make our state more responsible.

Sincerely,

[Your name]

READY TO ADVOCATE FOR ENERGY?

Use this sample letter as your guide when you're ready to write to your legislator to seek support for energy audits of all schools in your area.

Sample Session 7

Gearing Up to Go

AT A GLANCE

Goal: The girls brainstorm for their Innovate project.

- Opening Ceremony: Risk and Impact!
- How Are Those Plants Doing? And the Team?
- Lightbulbs!
- Thinking About a Team Choice!
- Option: Old-Fashioned Silhouettes
- Closing Ceremony
- Looking Ahead to Session 8

MATERIALS

- **Opening Ceremony:** Investigate awards (optional).
- **Old-Fashioned Silhouettes:** Light (use brightest bulbs from lightbulb activity); white paper or recycled paper: black paper; pencils, scissors, and glue sticks.
- **Lightbulbs!:** Photocopies of page 79, or make your own on reused paper.

GREEN THINGS

If the girls are still investigating buildings, let them come up with their own opening ceremony!

If they're stalled, suggest they go around their circle naming as many green things as they can think of—peas, lettuce, green thumb, geckos, fried green tomatoes, dollar bills.

The girls might like knowing that the color green became associated with environmentalism in early 1970s in West Germany with Green Campaign for the Future, which grew from protests against nuclear power plants.

AS GIRLS ARRIVE

Invite them to check out the profile of architect Sarah Susanka (page 67 in the girls' book), whose teacher gave her a puzzle—and a lesson in thinking outside the box. If the girls need help solving the puzzle, here's the answer:

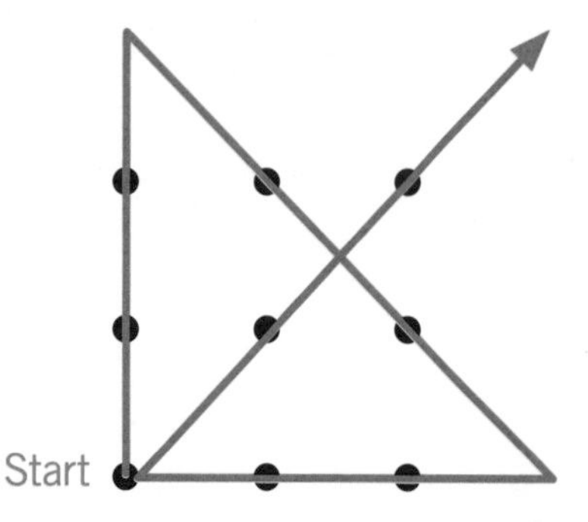

Opening Ceremony: Risk and Impact

Start off by acknowledging all that the girls have learned so far. Invite them to check the steps to the Investigate Award on page 107 of their book. If they've completed them, they can now earn the award. Encourage the girls to think about all the challenges they've taken on, and what they've learned as a result. Here are some guiding questions for the discussion:

- *What new things have you learned and thought about while investigating?*
- *What's it like to take a positive risk and do something new? Did you make any mistakes along the way? How did that make you feel?*
- *What other positive risks do you take in life? How do they help you? What have you learned from them?*
- *What happens if you make a mistake?*

Congratulate the girls for all they accomplished as leaders who investigate! Ask each girl to say something about the impact their investigation efforts might have on the environment. Here are some "fill in the blank" examples that might jumpstart their thinking. You might write one on a large piece of paper or chalkboard to help the girls focus.

- We suggested ______________________ to the building manager and that helps Earth because ______________________.
- Now that we know ________________, we can keep on _____________.
- If every building we investigate becomes more energy efficient, Earth would ____.
- We educated and inspired ______________________ by speaking up about ______________________.
- We advocated for _____________________ by ______________________.
- How Are Those Plants Doing? And the Team?

GROUP DISCUSSION TIPS

Give each girl the opportunity to voice her opinion. Some girls may be more assertive than others. Encourage each girl to speak up.

Invite the girls to gather the plants together and remove the bag(s) from the test plant(s). Ask the girls if they can tell which plant is the control plant and which are the test plant(s). By now, it should be obvious that the control plant is healthy and has flourished, while the ones deprived of light are less healthy. Ask again, *What you can conclude about the importance of light to a plant's growth?* Then ask, *How does food help you live and grow? Do you know how you synthesize the energy from food? What would happen if you didn't get the energy you need? How would you feel?*

Then say: *Speaking of energy, this is a good time to check in on our team energy. How is everyone getting along? How is everyone doing with sharing, helping, etc.? Have any conflicts arisen? How are they being settled?*

Lightbulbs!

Get a discussion going about what being innovative really means. You might ask: *When have you been innovative in your life? And on this journey? How has our team been innovative together?*

Then use the "lightbulbs" on the next page to inspire the girls to start thinking about their own Innovate effort. Each one offers a fine example of "innovation," "positive risk-taking," or "acting on values to create lasting change" for the environment.

Place the cut-up bulbs, or your own versions on slips of recycled paper, in a bowl or bag. Invite small teams of girls to choose a lightbulb, turn to the example in their book, and then respond to the questions on the lightbulb.

Then gather the full team together again to share their answers.

Thinking About a Team Choice

Transition the girls to brainstorm about possible Innovate projects. Using their "lightbulbs" and everything else they've learned along the journey, guide them to make one big list of all their ideas. Encourage them to be open and alert between now and their next gathering. You might say: *What other "lightbulbs" do you see around you, at school, in your community?*

If the girls have lined up some energy experts as visitors to their gatherings, the girls can make use of the Interview tips on page 81 of this guide to help them get the most useful information from their guests.

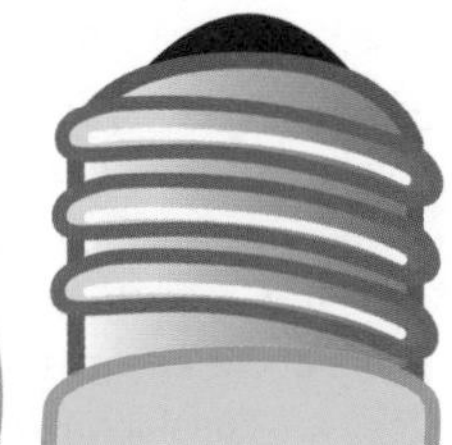

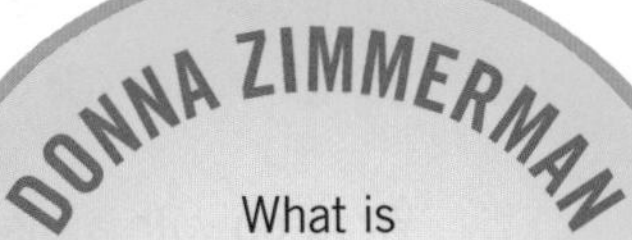

JENNI LARMORE

What was innovative about Jenni's idea? What positive risks did she take?

page 24

WHAT CAN YOU DO?

What did you learn from your own Energy Pledge? What positive risk did you take? What values from the Girl Scout Law did you use?

page 18–20

DONNA ZIMMERMAN

What is innovative about Donna's work? What risks does she take running her own business?

page 22

ABBE HAMILTON

What values is Abbe living? (Think about all the lines of the Girl Scout Law!) How does her project create lasting changes on Earth?

page 34

NANCY JUDD

How does Nancy educate and inspire other people? And Dez? What can you learn from her? Why is it important to educate and inspire other people to care about Earth?

pages 42–43

ROBIN CHASE

What is innovative about Robin's project? How will her project create lasting change?

page 84

¡VAMOS YA!

Why don't the girls just give up when they have a hard time getting their "walking bus" started? What challenges do they face? What positive risks are they taking?

pages 98–99

RECYCLE CINDY

What's innovative about what Cindy does? How does her work inspire other people?

page 41

TALK WITH EXPERTS!

Encourage the girls to think about energy experts who might visit their next team gathering. The girls can interview them to learn more about energy topics they are considering for their Innovate project.

Share the "How to Conduct an Interview" tips on page 81, and encourage the girls to use them when talking with their guest.

Let the girls know that before they choose an Innovate project, they have a chance to consider many ideas: all the ones in their books, plus any that come to mind, even carpools (including ones to Girl Scout gatherings!) and no-idle zones for cars at school.

Wrap up with a discussion about team decision-making. You might ask:

What happens in your life when you are part of a group and there is a conflict? How do you resolve things? Let's brainstorm a list of good techniques for conflict resolution.

Here are some techniques the girls might consider:

- Listen to all options.
- Give everyone a chance to speak.
- Keep talking.
- Compromise.
- Get help from a trusted adult.
- Vote (but will those whose votes don't win have their feelings hurt?).

Option: Old-Fashioned Silhouettes

Explain to the girls that a silhouette is a traced outline of a person's face in profile. It looks like a black shadow on a white background. Silhouettes were a way for people to create likenesses of loved ones before photography was invented. Then get started, giving each girl a chance to create a silhouette.

- Invite the girls to pair up, and then turn down the lights in the room.
- Have one girl sit in a chair in front of a sheet of paper that's been taped to the wall.
- Place a bright light so that the girl's shadow appears on the paper. Have the other girl trace the outline of the shadow with a pencil.
- Encourage the girls to sketch details, like eyelashes and stray locks of hair. Then put up another piece of paper so the girls can switch places.
- Each girl can dab the glue stick lightly onto the white paper with her silhouette and paste it onto a sheet of black paper, and cut around the outline. Then they can pull off the white paper and stick the black one onto another sheet of white paper to make their finished silhouette.

MAKING A GIFT OF IT

The girls might like to make silhouettes of friends or family members as a record of this moment in time—and this energy journey! They can use the flip side of used pieces of paper, or their recycled paper.

Encourage the girls to sign and date their creations.

How to Conduct an Interview

Try practicing these tips on friends and family ahead of meeting with your energy experts:

1. Think about what you hope to accomplish. What kind of information do you need?

2. Next, think about whom to interview. Who's an expert in this field? Who can best answer the questions you have?

3. Set up an interview. Call or e-mail the person, explain who you are and what you're doing, and ask if she will speak to you. Arrange a time and a place.

4. Do some research beforehand so you know some things about the person and her area of expertise. (This will give you confidence!)

5. Now, what are you still curious about? Prepare a list of questions. Think about the flow of the interview. Ask simple questions first, then move on to more complex ones. Group questions about one subject together. Then move on to the next subject.

6. When you meet, introduce yourself. If you want to record the interview, ask permission. Assure the person that you're recording so that you will have accurate quotes. But still, take careful notes. You never know when your recorder might fail.

7. Listen closely to what the person says. If you don't understand something, ask her to repeat the answer, or ask for clarification. Check details, spellings, dates, etc., as you go along.

Remember, you don't have to follow your list of questions word for word. If the person says something that surprises you or makes you curious, ask a follow-up question. Sometimes that's how you get the best information.

8. Try not to get so caught up in your note-taking that you don't make eye contact with the person. Nod your head when you're really interested in what is being said.

9. Near the end of the interview, quickly scan your list of questions to make sure you've covered what you need to know.

10. At the end of the interview, ask the person if there's anything she wants to add. Then thank her for her time and for sharing her expertise. Then tell her, "I'll be going over my notes. If I have other questions or need to clarify anything, may I contact you again?" If you're uncertain about anything, check back.

WHAT WOULD DEZ EAT?

Ask the girls to imagine that Dez is not a spider! Invite them to plan a day of energizing foods for her! What slogans can they come up with for what Dez eats?

Energizing Food

Check in and see what new ideas girls have about making "energizing food choices." You might ask:

- *Since you started paying attention to what advertisements tell you about food, have you changed your mind and made choices based on energizing food?*
- *What new snacks have you enjoyed trying along the journey?*
- *How does it feel to make a choice based on what you know truly makes you feel energized?*

Closing Ceremony

Get out the boom box or portable stereo. Explain that we sometimes take energy so much for granted, we only think about it when it stops. Say something like: *Let's think about energy and have fun with energy by playing a freeze dance game. It's simple. Everybody dances in whatever style they want. When the music stops, the dancers freeze. Whoever is still moving is out.*

Invite a girl to volunteer to be the monitor who starts and stops the music and watches to see who's out. Then ask the rest of the girls to stand up and begin playing. The last one still in is the winner.

Looking Ahead to Session 8 (and Possibly 9!)

At their next gathering, the girls will check out the walkability or bikeability of their community. To determine how safe and pleasant a walking or biking route is, they'll complete a walkability/bikeability survey. This will get them outdoors in an active way. It will also be a great step toward an Innovate project. (Walking school bus? Bike Train? A new idea from the girls?)

It's best to walk the chosen route in the daylight hours. The girls might consider mapping a route that starts at and/or includes the homes of one or more of the girls in the group and ends at the team's gathering place. That way, all the girls can meet at one home, and then everyone will walk to the meeting place together and collect data along the way. Make copies of the Walkability and Bikeability Survey (pages 86–87) and familiarize yourself with the traffic problems and other issues you and the girls might encounter.

TAP THE NETWORK

Ask for adult volunteers to join you in accompanying the girls on any outings for their Innovate project—make sure there's enough supervision. You might also ask volunteers to provide a handy "walkable" snack.

Another option: The team might visit a hybrid car dealer or have a hybrid owner visit with her car to show the girls how it works. Call or e-mail an auto dealer in your area to arrange a visit and/or put your Network into action to find a hybrid car owner to talk with the girls.

Planning Time: Innovate!

Our Innovate idea is:

We will carry out our project on these dates and times:

Our project will make a difference because:

Specific steps we will take:

1.

2.

3.

4.

5.

We will talk to these people:

We will ask others to get involved by:

We will get their attention by:

We will use our creativity to:

We will take a positive risk by:

When we earn our Innovate Award, we'll be really proud that we:

Sample Session 8

Moving in New Directions

AT A GLANCE

Goal: The girls explore the energy spent and saved in getting from here to there as they continue to move toward a team decision on an Innovate project.

- **Walkability/Bikeability**
- **Making a Team Decision on an Innovate Project**
- **Next Steps**
- **Closing Ceremony**

MATERIALS

- **Walkability/Bikeability:** copies of survey, page 86-87.
- **Making a Team Decision:** Innovate project idea poster board; markers.
- Energizing snack of the girls' choice.

PREPARE AHEAD

Have copies of the Walkability and Bikeability Survey ready for the girls and secure some extra Network volunteers willing to join the outing.

Opening Ceremony

The girls can go around the circle and say the name of one star, constellation, or planet and how they first learned of it. Alternately, they might name their favorite walking or biking experience and say why it was special.

Walkability/Bikeability

Before the team sets out to review their proposed walking or biking route, engage the girls in a discussion about walking and biking to school. Ask: *When was the last time you walked or biked to school? Would you consider doing it more often? What would it take to get you doing it more often?*

Also ask: *What makes a neighborhood safe and pleasant to walk in? What about sidewalks, traffic, streets, big dogs?* Suggest that they keep an eye out for those things as they walk. Also suggest that they decide how to walk in an orderly and safe fashion. Two-person teams? Pass out copies of the survey (pages 86–87) to the teams, and suggest they make notes as they walk.

When the group reaches their meeting place, the girls can complete the survey together. You might ask: *Did the walk energize you? Did you find out anything new about the neighborhood or about your sister Juniors?*

Making a Team Decision on an Innovate Project

Gather the girls around and bring out the list of Innovate project ideas. Ask if anyone has any new ideas to add. Then, go down the list and ask each girl to discuss her favorite ideas. As the discussion builds and girls point out pros and cons of various ideas, guide the group to narrow down the project choices. Does any idea emerge as a clear front-runner? If yes, ask the group, "Do we all agree that this is a good choice?" If not, keep the discussion going. If still nothing emerges as a favorite, consider taking a vote to narrow down the ideas.

Congratulate the girls on making a commitment to innovate as a team. Remind them that leaders innovate!

PICKING A PROJECT

A project doesn't have to be big to be worthwhile. Small, well focused projects can have great impact and create lasting, sustainable outcomes. A successful project is one that helps the girls' develop their skills and talents—and encourages them to be smart about energy while emerging as leaders!

Next Steps

Engage the girls in a discussion of how they'd like to *GET MOVING!* on their Innovate project. Make use of the Planning Time: Innovate! worksheet (page 83) and Innovate Checklist (page 90–91).

DECISIONS MADE EASY

If the girls have already journeyed through *Agent of Change*, remind them about using "Fist-to-Five" to make decisions that everyone feels good about.

Closing Ceremony

Invite each girl to offer one wish for the team's Innovate project and also give an example of something she can do to make that wish come true.

Walkability and Bikeability Survey

Answer the following questions using this rating scale:

1 (many problems) **2** (some problems) **3** (good) **4** (very good) **5** (excellent)

1. Is there room to walk/bike? Consider the following:

- Are sidewalks or paths available the entire way?
- Is there enough room for two people to walk side by side?
- Are the sidewalks broken or cracked? Is the bike path free of potholes or dangerous drain gates, slippery surfaces when wet, or bumpy or angled railroad tracks?
- Are the sidewalks or bike paths blocked by poles, signs, shrubbery, dumpsters, cars, or other obstacles?
- How's the traffic? Was there too much traffic? Are the sidewalks too close to fast-moving traffic?
- Are there other problems?

Rating: (circle one) **1 2 3 4 5**

2. Is it easy to cross streets? Consider the following:

- Are the streets too wide?
- Do the traffic signals make your group wait too long or not give you enough time to cross?
- Do the streets need traffic signals? Are the crosswalks marked?
- Do parked cars block your view of traffic? Do trees or plants block your view of traffic?
- Are there curb ramps or do the curb ramps need repair?

Are there other problems?

Rating: (circle one) **1 2 3 4 5**

3. Do drivers behave well?

- Do drivers look before they back out of driveways?
- Do they yield to pedestrians crossing the street?
- Do they drive at safe speeds? Or do they drive too fast? Do they speed up to make it through traffic lights?
- Are there other problems?

Rating: (circle one) **1 2 3 4 5**

4. Is it easy to follow safety rules? Can your group . . .

- Cross at crosswalks, or where you can see and be seen by drivers?
- Stop and look left, right, and then left again before crossing streets?
- Walk on sidewalks or shoulders facing traffic where there are no sidewalks?
- Cross with the light?

Rating: (circle one) **1 2 3 4 5 6**

5. Is the walk pleasant?

- Is there a need for more grass, flowers, or trees?
- Are there scary dogs or scary people?
- Is the lighting good? Is there lots of litter or trash?
- Is there dirty air from automobile exhaust?
- Are there other problems?

Rating: (circle one) **1 2 3 4 5**

How walkable is your route? Add up your ratings and decide.

21–25 points: Fantastic! Your route is great for walking. Talk to your school to get permission to make it official. Start educating others about the benefits of a walking bus, and inspire them to join in!

16–20 points: Nice! Your route is pretty good. Who can you reach out to so that it can be even better before you try to go official with a walking bus?

10–15 points: OK, but your route could be a lot better. What local officials might you reach out to so that improvements can be made that will allow you to pursue this as a walking/biking bus route? Who can you inspire to join you in seeking official improvements?

Below 10 points: Time to brainstorm a new Innovate project!

** Adapted with permission from the National Center for Safe Routes to School (saferoutesinfo.org).*

Sample Sessions 9 & 10

Innovate!

Your team has been on the move, exploring personal energy use, the energy in buildings, and the energy used in moving around. Now is their opportunity to carry out an Innovate project that makes use of something they've learned along the journey to create positive change in the community.

AT A GLANCE

Goal: The girls plan and carry out their Innovate project, taking action to create changes in energy use on Earth and educating and inspiring others along the way.

- **Opening (and Closing) Ceremony Ideas**
- **Teamwork and Conflicts**
- **Innovate Checklist**

MATERIALS

- **Innovate Checklist** and whatever materials the girls need to carry out their Innovate project.

PREPARE AHEAD

Use the checklist, tips, and options provided here as your team needs them. As their work progresses, the team may change course. The project may shrink or grow, or even turn into something different altogether. Don't worry. What matters is that the girls are using their energy, working as a team, and building leadership skills.

Let the girls decide whether their final celebration will be a formal ceremony, an energetic party, an outing, or a gathering that includes Brownies who will be on the journey next. Don't forget to be mindful of energy use! Encourage the girls to make the invitations from their recycled paper!

Opening (and Closing) Ceremony Ideas

Even if the team is out and about, take a few minutes to huddle and remember that this time spent together, acting on behalf of Earth, is important and special! Ceremonies offer an opportunity for the girls to take pride in how they're learning and growing! The Juniors might try one of these ideas (or one of their own!). Form a circle and:

- Say one word that represents how you feel about your Innovate effort.
- Share one new idea you have gotten during the *Get Moving!* journey.
- On a sheet of (recycled or homemade!) paper, write the names of everyone who has helped you and what you learned from them.
- Dez says, ________________________ (add your own quippy line!)

Teamwork and Conflicts

Planning and carrying out the Innovate project is a great way for girls to work as a team and resolve conflicts that may develop along the way. Who "gets to talk" at a presentation? Who didn't speak up? Whose idea was forgotten? Who forgot to do her share?

All of these issues can be great learning experiences for Juniors, if you coach the girls through them! Here are some tips to keep things moving smoothly:

Celebrate the positives: Perhaps each girl can "gift" a team member by naming something she is really good at ("I appreciate that Leticia always listens to everyone").

Take team time outs: Encourage girls to call "time," or call it yourself if they don't. Everyone stops focusing on "doing" and thinks instead about how they are interacting with each other. Would it help to:

- Listen to new ideas and adjust the plan?
- Ask each other how we are feeling and what it would take to feel better?
- Find a compromise on something?
- Trade places/switch roles?
- Make a team agreement about how we all promise to act?
- Consider what the Girl Scout Law advises us to do?

As team work progresses, encourage girls to make a joint "Top Tips for Resolving Conflicts" list. Where in their lives could they use some of these?

KEEP THE ENERGY HIGH

The project may not go exactly as planned! The girls could reach out and talk to others about an idea and find that it is not possible for a lot of reasons! Encourage girls and praise them for trying! What did they learn from their efforts? What can they adjust? Your goal is to give girls a springboard to a lifetime of trying to make a difference. So, keep the enthusiasm up, even as plans shift!

Innovate Checklist

As the volunteer, how do you know if the Innovate project is a good experience for girls? Use this handy checklist. Does the project meet these checkpoints? Use the checkpoints as a base for guiding the girls through their planning process, too.

CHECKPOINT	ACCOMPLISHED BY
Brainstorm. Girls use ideas from their book and from all they've learned and accomplished on the journey. Will they start a walking club? A bike program? Advocate for carpools? Create no-idle zones? Organize families to make energy pledges?	
Team Decision. The girls decide on their Innovate project together as a team.	
Doable Plan. With guidance and coaching from you and their Friends and Family Network, the girls make a practical plan based on the time and resources available to them.	
Keep It Going. Guide girls to strive toward lasting change. Who can they involve so the effort keeps going? (Instead of a one-day event, how about asking the PTA to aim for a monthly event?)	

CHECKPOINT	ACCOMPLISHED BY
Learn Some More. Girls have a chance to meet and interview a few (two to earn Innovate!) people who can give them ideas about their project. Work the Friends and Family Network to identify people who are "in the know" about the focus of the girls' project.	
Educate and Inspire. Girls have a chance to ask others to get involved! Who can girls reach out to? What are their creative ideas for inspiring others? Can they ask others to make a pledge? Set up a display at the library or school? Present to the Friends and Family Network? Make their own comic story and share it with other girls?	
Speak Out. The girls have opportunities to ask for what they need to make change. Can they talk to the school principal about bike storage? Ask families not to idle cars at the curb? Speak to local businesses about turning lights out at night? Make sure each girl gets a chance to ask someone to do something!	
Team Up. The girls have a chance to observe and improve their teamwork as they Innovate.	
Reflect. Guide the girls to take pride in all they have learned and accomplished! This could be part of their closing celebration.	
Give Thanks! The girls thank everyone who has assisted them along the journey. (How about notes on recycled paper?)	

Sample Session 11

Crossing the Finish Line

AT A GLANCE

Goal: Girls reflect on and celebrate their accomplishments along the journey.

- **Opening Ceremony: Reflecting on the Journey and a Human Perpetual Motion Machine**
- **Innovate Award Ceremony**
- **Celebrate!**
- **Getting Energized About the Future**

MATERIALS

- **Opening ceremony:** index cards and ball of yarn or string, or name tags.
- **Innovate Award Ceremony:** Innovate Awards.
- **Celebrate:** Team strand of paper beads; boom box for freeze dance, popcorn lightbulbs or another snack.

Opening Ceremony: Reflecting on the Journey and a Human Perpetual Motion Machine

Gather the girls in a circle and invite them to reflect on all they've accomplished along the journey. By now, they've had a chance to think about the questions in their Energy Tracker, and they might want to share their thoughts with the team, or with the team and their guests:

- *What have you discovered about who you are and the values you stand for?*
- *How have you connected with others on your Junior team and beyond?*
- *How else can you use your energy to Innovate in big and small ways to Take Action and make the world a better place?*

Now invite the girls and their guests to build a girl-designed, girl-powered Human Perpetual Motion Machine in which they are the gears, pistons, and other moving parts that make the machine go.

You might say, *As we have realized along our journey, leaders are full of energy. They know how to energize everyone around them to team up and do things that benefit Earth and people. Let's use our mighty, unstoppable, renewable energy and imagination to make a Human Perpetual Motion Machine go!*

First, ask the group to name forms of renewable energy, ones that don't contribute to global warming, such as solar, wind, or water power. Ask the group to pick one to power their machine. Then ask for a pair of volunteers to be this energy—and to make a name tag or card for themselves (using the materials provided) that states or symbolizes it. Then ask them to practice how they will act out this energy and "start" the machine everyone else will make.

Now ask everyone else to think of a part of a machine—gears, levers, crankshaft, spark plugs, wheels, pulleys, etc.—and to write their choice on a name tag or an index card, too. (It's OK if more than one person chooses the same part. There can be more than one of each.)

Then distribute the tags or cards randomly, or put them in a pile and ask everyone to pick one.

Invite the girls and their guests to form a circle. Ask them to imagine what their Human Perpetual Motion Machine will accomplish. Perhaps it will compact a trash heap into a brick? Produce enough energy to grow food for everyone on Earth? Fuel a new metrorail through town so we can give up our cars? Encourage the group to get creative! What's an energy solution the community needs? The Human Perpetual Motion Machine can provide it!

Once everyone can imagine what the machine accomplishes, invite the first girl to play the form of renewable energy that was first chosen. She starts the machine by "flipping the switch" or "plugging in the plug." (She can lift the arm of the next person and pull it down or pretend to "plug" the arm into a socket.) Once that person is powered, she touches the next person, who performs the action of her engine part (a piston jumps up and down; a crankshaft moves backward and forward, a wheel spins round and round—the more energetic the better!). And that girl then touches the next person, who performs her action, and so on around the circle.

When all the engine parts have jumped, spun, rumbled, hummed, etc., the last person falls down, and the wind or other energy source picks her up, and starts the machine again. Encourage the group to make the next round even more energetic than the first!

When the girls and their guests have enjoyed a few rounds, wrap up by inviting the girls to reflect on all the energy smarts and leadership skills they've developed as they've traveled along this journey—and all that they want to accomplish going forward, no matter how challenging. Great leaders are positive risk-takers who motivate others! You might ask:

- *What are you most proud of accomplishing during the energy journey?*
- *What do you think you'll remember most?*
- *What have you learned that you'll take along with you?*
- *Is there a way to keep the momentum going?*

Point them to the energy pledge on the inside back cover of the girl book, and let them know that their commitment to using energy wisely doesn't have to end with this journey. Lasting change goes on and on!

Innovate Award Ceremony

Hand out the Innovate awards, and . . .

. . . Celebrate!

The girls might choose from these energizing ideas, or come up with their own way to celebrate:

- **Invite a Brownie group** that plans to take this energy journey next year. "Meet and mingle" and share favorite moments of the journey with the younger girls to get them excited about the adventures to come. Let them know what challenged or surprised you the most. Give your team strand of recycled paper beads—or a few of its beads—as a gift to the Brownies, and invite them to add to the strand next year to keep the energy flowing!

- **Play a giant game of Freeze Dance** so they can enjoy seeing their guests cutting loose with their energy. Get out the boom box and invite everyone to *Get Moving!*
- **Create a team photo** with everyone wearing their paper bead necklaces.
- **Sign one another's books** and add an energizing memory, too.
- **Share with their guests** the artwork, keepsakes, and other items they've created and collected along the journey.
- **Share their energy pledges** and invite their guests to write and commit to their own pledge.
- **Create an energizing snack,** such as popcorn lightbulbs or anything else the girls enjoyed along the journey.

Getting Energized About the Future

If the Junior team will continue on with other Girl Scout adventures this year, spend a few moments at the conclusion of the celebration getting the girls excited about what's ahead.

If this concludes the year's Girl Scout experiences for these Juniors, you can still engage the girls in thinking about how they might participate in Girl Scouts next year. A new journey perhaps? Camping? Other events going on through your Girl Scout council? Be sure the girls (and their Friends and Family Network) know how they can keep their Girl Scout adventures growing.

Now, Take a Moment for Yourself!

Congratulate yourself for an energizing journey! Also, look back at your thoughts on the three keys to leadership on page 27.

As you've guided the Juniors to some new discoveries about themselves, connections with one another and the community, and opportunities to find ways to Take Action, what have you learned about the keys to leadership?

What have you discovered about acting on your values?

What new connections in the community have you made? What do they mean to you?

As you guided the girls to Take Action for the environment how have you contributed to making the world a better place?

See you've really been on the move! Take a break! Celebrate! Then remember: There's a whole new Girl Scout journey waiting for you!